GREAT DISASTERS

THE BLACK DEATH

Other books in the Great Disasters series:

Earthquakes

Tornadoes

Volcanoes

GREAT DISASTERS

THE BLACK DEATH

Jordan McMullin, *Book Editor*

Daniel Leone, *President*
Bonnie Szumski, *Publisher*
Scott Barbour, *Managing Editor*

San Diego • Detroit • New York • San Francisco • Cleveland
New Haven, Conn. • Waterville, Maine • London • Munich

For more information, contact
Greenhaven Press
27500 Drake Rd.
Farmington Hills, MI 48331-3535
Or you can visit our Internet site at http://www.gale.com

Cover credit: © The Louvre, Paris/Bridgeman Art Library, New York

Dover Publications, 79
Library of Congress, 67
National Library of Medicine, 21, 45, 50

LIBRARY OF CONGRESS CATALOGING-IN-PUBLICATION DATA

The Black death / Jordan McMullin, book editor.
p. cm. — (Great disasters)
Includes bibliographical references and index.
ISBN 0-7377-1498-0 (lib. : alk. paper) — ISBN 0-7377-1499-9 (pbk. : alk. paper)
1. Black death—Popular works. I. McMullin, Jordan. II. Great disasters (Greenhaven Press)
RC171 .B535 2003
614.5'732'00994—dc21 2002034727

Printed in the United States of America

CONTENTS

Foreword 8

Introduction: The Plague to End All Humankind 10

Chapter 1: "We All Fall Down": The Causes and Immediate Devastation of the Black Death

1. The Arrival of the Plague in Europe
by Charles L. Mee Jr. 16
In 1347, the *Yersinia pestis* bacteria moved by merchant ships from the East to Italy, thereby entering all of Europe. The explanations and suggestions of physicians did little to help the afflicted.

2. Recent Scientific Reassessments of the Black Death
by Norman F. Cantor 26
Although most biologists take for granted that the Black Death was caused primarily by the bubonic plague, reassessments of historical evidence suggest the possibility of another disease operating at the time, including anthrax.

Chapter 2: "The Whole Place Was a Sepulchre": Contemporary Chronicles of the Black Death

1. Death Comes to Florence
by Giovanni Boccaccio 33
In the preface to *The Decameron*, Boccaccio describes the horrors visited upon the city of Florence when the pestilence arrives.

2. The Impact of the Black Death in England
by Henry Knighton 41
The Black Death caused social and economic chaos in England. The aristocracy tried in vain to suppress the economic repercussions of the devastation.

3. A Treatise on the Prevention and Cure of the Plague
by John of Burgundy 47
A contemporary medical treatise suggests various ways to avoid contracting the plague. These include the inhalation of herbal concoctions, and the avoidance of hot foods, baths, and sexual intercourse.

Chapter 3: "The World Turned Upside Down": Reactions to and Repercussions of the Black Death

1. Desperate Responses: The Flagellants and the Persecution of Jews
by Philip Ziegler 55
The Flagellants were a Christian religious order that moved from town to town, scourging their own flesh in the hopes that self-punishment would please God and the plague would end. Many Christians blamed Jewish communities for the plague, massacring them in vast numbers.

2. The Devastating and Far-Reaching Effects of the Black Death
by Robert S. Gottfried 70
After the massive devastation of the plague, religion, human behavior, and the arts were characterized by pessimism and a general preoccupation with morbidity and death.

3. Advances in Medicine, Surgery, and Public Sanitation
by Anna Montgomery Campbell 88
Because of the failure of medieval medicine to combat the plague, the pope and other public officials finally allowed for the dissection of human corpses and the serious study of anatomy.

4. Competing Historical Interpretations of the Black Death
by Faye Marie Getz 94
With the rise of epidemiology in the nineteenth

century, historical interpretations of the Black Death tended to be "gothic" in nature—that is, scholars believed that the horrific disaster actually had a silver lining, since it ushered in a new era of social change.

For Further Research 103

Index 107

FOREWORD

Humans have an ambivalent relationship with their home planet, nurtured on the one hand by Earth's bounty but devastated on the other hand by its catastrophic natural disasters. While these events are the results of the natural processes of Earth, their consequences for humans frequently include the disastrous destruction of lives and property. For example, when the volcanic island of Krakatau exploded in 1883, the eruption generated vast seismic sea waves called tsunamis that killed about thirty-six thousand people in Indonesia. In a single twenty-four-hour period in the United States in 1974, at least 148 tornadoes carved paths of death and destruction across thirteen states. In 1976, an earthquake completely destroyed the industrial city of Tangshan, China, killing more than 250,000 residents.

Some natural disasters have gone beyond relatively localized destruction to completely alter the course of human history. Archaeological evidence suggests that one of the greatest natural disasters in world history happened in A.D. 535, when an Indonesian "supervolcano" exploded near the same site where Krakatau arose later. The dust and debris from this gigantic eruption blocked the light and heat of the sun for eighteen months, radically altering weather patterns around the world and causing crop failure in Asia and the Middle East. Rodent populations increased with the weather changes, causing an epidemic of bubonic plague that decimated entire populations in Africa and Europe. The most powerful volcanic eruption in recorded human history also happened in Indonesia. When the volcano Tambora erupted in 1815, it ejected an estimated 1.7 million tons of debris in an explosion that was heard more than a thousand miles away and that continued to rumble for three months. Atmospheric dust from the eruption blocked much of the sun's heat, producing what was called "the year without summer" and creating worldwide climatic havoc, starvation, and disease.

As these examples illustrate, natural disasters can have as much impact on human societies as the bloodiest wars and most chaotic political revolutions. Therefore, they are as worthy of study as the major events of world history. As with the study of social and political events, the exploration of natural disasters can illuminate the causes of these catastrophes and target the lessons learned about how to mitigate and prevent the loss of life when disaster strikes again. By examining these events and the forces behind them, the Greenhaven Press Great Disasters series is designed to help students better understand such cataclysmic events. Each anthology in the series focuses on a specific type of natural disaster or a particular disastrous event in history. An introductory essay provides a general overview of the subject of the anthology, placing natural disasters in historical and scientific context. The essays that follow, written by specialists in the field, researchers, journalists, witnesses, and scientists, explore the science and nature of natural disasters, describing particular disasters in detail and discussing related issues, such as predicting, averting, or managing disasters. To aid the reader in choosing appropriate material, each essay is preceded by a concise summary of its content and biographical information about its author.

In addition, each volume contains extensive material to help the student researcher. An annotated table of contents and a comprehensive index help readers quickly locate particular subjects of interest. To guide students in further research, each volume features an extensive bibliography including books, periodicals, and related Internet websites. Finally, appendixes provide glossaries of terms, tables of measurements, chronological charts of major disasters, and related materials. With its many useful features, the Greenhaven Press Great Disasters series offers students a fascinating and awe-inspiring look at the deadly power of Earth's natural forces and their catastrophic impact on humans.

INTRODUCTION

The Plague to End All Humankind

In recent times, as depleting natural resources alarm the world's citizens, and as overdevelopment and pollution concern a growing number of environmental advocates, it has become ever more important to understand the delicate balance between humans and their natural surroundings. In fact, although many might believe that the advent of the twenty-first century has incited a new awareness of the frequent collision between the earth and its human inhabitants, nature has been fighting back against human invasion since we first stepped out of a puddle of single-celled organisms. In his book *Plagues and Peoples*, William H. McNeill writes, "Everyone can surely agree that a fuller comprehension of humanity's ever-changing place in the balance of nature ought to be part of our understanding of history, and no one can doubt that the role of infectious diseases in that natural balance has been and remains of key importance."[1] Undoubtedly, in the battle of nature against humans, bacterial plagues have changed the world's course of events since historians first began recording their mysterious occurrences, but none has rivaled the immediate and long-term devastation of the Black Death of 1347 and 1348.

Modern medicine, with its arsenal of antibiotics, is now able to fight outbreaks of the bubonic plague, but at the time of its first recorded incidence—in Babylonian writings from around 2000 B.C.—humans did not understand the causes of or cures for such a bacterial infection, let alone understand that it was carried by fleas on the backs of rats infected with the *Yersinia pestis* bacillus. The Greek historian Thucydides describes in his *History of the Peloponnesian War* a plague outbreak that hit Athens in 430 B.C., and although it is impossible to determine that it was bubonic plague, the symptoms Thucydides describes seem to fit the bill.

Over the next one thousand years, the plague hit numerous cities in Asia and Europe, particularly Rome, but none of these outbreaks spread with the rapidity of the sixth century's Plague of Justinian. Although the origins of the Plague of Justinian are unknown, records show that it hit Egypt in the year 542. It went on to devastate the whole of the Roman Empire, then ruled by Emperor Justinian. As would be the case with the Black Death nearly eight hundred years later, the Plague of Justinian metamorphosed from an isolated epidemic—affecting only one small area—into a widespread disease of mass proportions, primarily because of the new prevalence of trading and trade routes. When merchants and tradesmen moved from city to city, so did the bacteria they were carrying; soon, the epidemic became a pandemic. In fact, the Black Death spread from Asia to almost all of Europe in only four years (1347–1351) because of improvements in travel and the availability of commerce routes in the mid–fourteenth century. Humans, in our infinite ignorance of the complicated mechanics of nature, can unknowingly turn a bad situation into a horrible one.

Europe hadn't seen the last of the Black Death. Another characteristic of a pandemic is that it returns every few years after the first major outbreak. This was the case with the Black Death and is one of the reasons it wreaked such consistent havoc on the population of Europe throughout the rest of the fourteenth century. After the major outbreak waned in 1351, it flourished again in the 1360s, then in 1370. (The only plague after the Black Death to equal its horror was the Great Plague of London in 1665—the last major plague to hit the British Isles to date.)

On the eve of the Black Death, Europe was already experiencing a number of economic and social crises. In fact, historians since the nineteenth century have debated whether the chaos and social upheaval "caused" by the Black Death had actually begun long before the pandemic. Indeed, some argue that the Black Death merely exacerbated the moral and social decline of Europe and its feudalistic system—a decline that had been set in

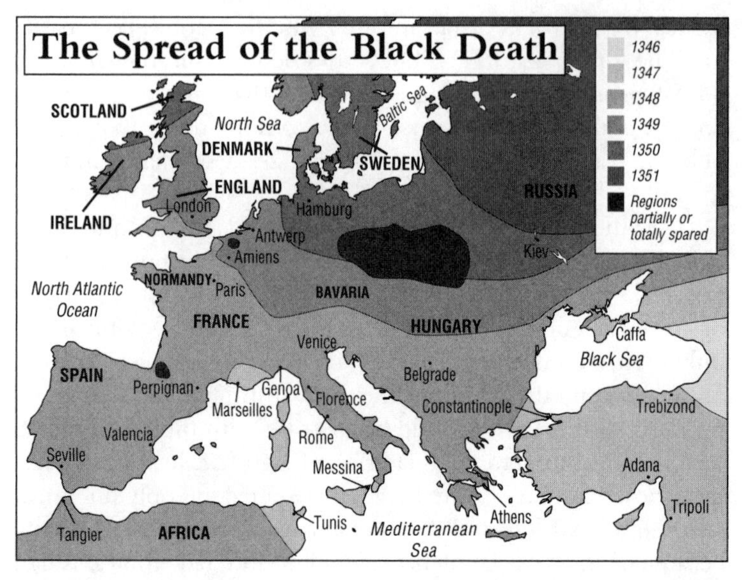

motion as early as the start of the thirteenth century. Still others argue that the Black Death wasn't the cause of anything at all. Rather, some say, it was an effect. According to this argument, the human conditions of overpopulation and a scarcity of resources kick-started the plague pandemic. Nature, after all, is capable of determining what it can sustain, and it maintains a precious ecological balance through mediums like disease epidemics. In any case, most historians agree that the social fabric of most of Europe was under great strain by the year 1347.

The tenth through the thirteenth centuries had been relatively prosperous; agricultural innovations, relatively few diseases, and a stable political system all helped to increase Europe's population from 25 million in 950 to 75 million in 1250. Medieval society operated under a system of "trifunctionality," a three-part division of class in which the clergy occupied the highest rank, the military and aristocrats sat right beneath, and everyone else—the farmers and merchants who supported the two classes above them—was assigned to the bottom rung. Agriculture was the heart of the European economy, but the peasants who worked the land did not have much freedom. They worked in a feudal, or "manorial," system, according to which they had to pay rent to a rich landowner in order to work his fields for their food, and most of what they produced went to the landowner for free. The system worked fairly well, particularly for the aristocrats who received free labor, and the overabundance of foodstuffs allowed a market economy to gain a foothold in late medieval society. New trade routes opened, especially to Asia and the Middle East, and there was a new emphasis on university education. All this prosperity came to an end, however, with the start of the "Little Ice Age" in the 1250s. The weather became colder and wetter, resulting in a series of poor harvests that then led to famines throughout the 1290s. Living standards dropped, tensions between the upper and lower classes increased, and the clergy were losing some of their power in society. In short, as author Robert S. Gottfried writes, "the bonds and structure of European society, intricately developed through 400 years, were coming apart."[2]

In his book *The Black Death*, Philip Ziegler writes that "One third of a country's population cannot be eliminated over a period of some two and a half years without considerable dislocation to its economy and social structure."[3] Indeed, Ziegler also believes that the Black Death had profound consequences on the "minds of its victims"—psychological scars that lasted into the long term, spanned generations, and forever changed people's understanding of the world and how they ran it. Many historians agree with Ziegler's sentiments; they argue that the Black Death's massive toll in lives caused a temporary social upheaval that had a profound impact on all of society's institutions, including the church, governments, and intellectual fields from medicine to mathematics. Even more far-reaching, some scholars believe that art and literature also became reflective of the Black Death, preoccupied with images of suffering and the macabre. But the historical interpretations of the consequences of the Black Death have changed over time. More recently, the idea that the fourteenth-century plague changed history for all time has been downplayed. After all, it is impossible to make that claim for any one event, however calamitous. History does not fit into the neat categories that scholars often impose, and to bolster this claim, many have been arguing that most medieval social institutions survived the pandemic relatively intact.

What can possibly explain this drastic split in interpretation? Paul Slack gives us a clue in his introduction to *Epidemics and Ideas*: "Different diseases exist, and different microorganisms affect their human hosts and human society in different ways. Yet epidemics are also themselves intellectual 'constructs' which, once formulated, have a history, vitality and resilience of their own."[4]

In other words, history books are created by particular individuals in a particular time who have certain unexamined biases in the way they interpret the historical evidence available to them and re-create the event in writing. History doesn't just happen and then jump onto the pages of a dusty textbook. People write history, and individuals are prone to hold certain beliefs over others, certain interpretations over others. One of the truly fascinating qualities of the Black Death as a historical event is its perpetuity. Even six hundred years after the fact, scholars in virtually every intellectual field debate about the Black Death and its impact on our lives today. From the scientific causes to the artistic consequences, practically the whole world has been engaged in a lively conversation about this interesting yet elusive event. In the collection of viewpoints that follows, one can glean a sense of the range of this debate surrounding the pandemic, as well as the fourteenth-century culture that endured it.

Notes

1. William H. McNeill, *Plagues and Peoples.* New York: Anchor Press, 1976, p. 5.
2. Robert S. Gottfried, *The Black Death: Natural and Human Disaster in Medieval Europe.* New York: Free Press, 1983, pp. 16–17.
3. Philip Ziegler, *The Black Death.* New York: Harper & Row, 1969, p. 232.
4. Paul Slack, "Introduction," *Epidemics and Ideas: Essays on the Historical Perception of Pestilence,* edited by Terence Ranger and Paul Slack. Cambridge, England: Cambridge University Press, 1992, p. 8.

Chapter 1

"We All Fall Down": The Causes and Immediate Devastation of the Black Death

The Arrival of the Plague in Europe

By Charles L. Mee Jr.

Charles L. Mee Jr. is known mostly as a playwright and the author of several nonfiction books on diplomatic and political issues. In the following excerpt from his article on the Black Death, which appeared in the Smithsonian, *Mee describes how the* Yersinia pestis *bacillus, thought by many to be the cause of the plague, moved from rats to fleas to humans in its search for a viable host. Mee provides a summary of the popular account of the arrival of the Black Death in Europe in 1347 from China and the Middle East. He goes on to describe in terrifying detail the death wrought by the plague, and how the pandemic infused a general sense of fear and hysteria throughout Italy—the first European country to be overtaken by the disease. Medieval medicine was inadequate to battle the plague; physicians and priests explained the Black Death to be a result of corrupted air, an unfortunate planetary alignment, or God's anger at the wickedness of humans. The suggested preventative and curative measures, therefore, did little to help the afflicted. Mee concludes with a description of the three types of plague believed to be in operation at the time: bubonic, pneumonic, and septocemic.*

In all likelihood, a flea riding on the hide of a black rat entered the Italian port of Messina in 1347, perhaps down a hawser tying a ship up at the dock. The flea had a gut full of the bacillus *Yersinia pestis.* The flea itself was hardly bigger than the letter "o" on this page, but it could carry several hundred thousand bacilli in its intestine.

The Culprits of the Plague

Scholars today cannot identify with certainty which species of flea (or rat) carried the plague. One candidate among the fleas is *Xenopsylla cheopis,* which looks like a deeply bent, bearded old

Charles L. Mee Jr., "How a Mysterious Disease Laid Low Europe's Masses," *Smithsonian*, February 1990, pp. 66–79.

man with six legs. It is slender and bristly, with almost no neck and no waist, so that it can slip easily through the forest of hair in which it lives. It is outfitted with a daggerlike proboscis for piercing the skin and sucking the blood of its host. And it is cunningly equipped to secrete a substance that prevents coagulation of the host's blood. Although *X. cheopis* can go for weeks without feeding, it will eat every day if it can, taking its blood warm.

One rat on which fleas feed, the black rat (*Rattus rattus*), also known as the house rat, roof rat or ship rat, is active mainly at night. A rat can fall 50 feet and land on its feet with no injury. It can scale a brick wall or climb up the inside of a pipe only an inch and a half in diameter. It can jump a distance of two feet straight up and four horizontally, and squeeze through a hole the size of a quarter. Black rats have been found still swimming days after their ship has sunk at sea.

A rat can gnaw its way through almost anything—paper, wood, bone, mortar, half-inch sheet metal. It gnaws constantly. Indeed, it *must* gnaw constantly. Its incisors grow four to five inches a year: if it were to stop gnawing, its lower incisors would eventually grow—as sometimes happens when a rat loses an opposing tooth—until the incisors push up into the rat's brain, killing it. It prefers grain, if possible, but also eats fish, eggs, fowl and meat—lambs, piglets and the flesh of helpless infants or adults. If nothing else is available, a rat will eat manure and drink urine.

Rats prefer to move no more than a hundred feet from their nests. But in severe drought or famine, rats can begin to move en masse for great distances, bringing with them any infections they happen to have picked up, infections that may be killing them but not killing them more rapidly than they breed.

The *Yersinia Pestis* Bacillus

Rats and mice harbor a number of infections that may cause diseases in human beings. A black rat can even tolerate a moderate amount of the ferocious *Yersinia pestis* bacillus in its system without noticeable ill effects. But bacilli breed even more extravagantly than fleas or rats, often in the millions. When a bacillus finally invades the rat's pulmonary or nervous system, it causes a horrible, often convulsive, death, passing on a lethal dose to the bloodsucking fleas that ride on the rat's hide.

When an afflicted rat dies, its body cools, so that the flea, highly sensitive to changes in temperature, will find another host.

The flea can, if need be, survive for weeks at a time without a rat host. It can take refuge anywhere, even in an abandoned rat's nest or a bale of cloth. A dying rat may liberate scores of rat fleas. More than that, a flea's intestine happens to provide ideal breeding conditions for the bacillus, which will eventually multiply so prodigiously as finally to block the gut of the flea entirely. Unable to feed or digest blood, the flea desperately seeks another host. But now, as it sucks blood, it spits some out at the same time. Each time the flea stops sucking for a moment, it is capable of pumping thousands of virulent bacilli back into its host. Thus bacilli are passed from rat to flea to rat, contained, ordinarily, within a closed community.

For millions of years, there has been a reservoir of *Yersinia pestis* living as a permanently settled parasite—passed back and forth among fleas and rodents in warm, moist nests—in the wild rodent colonies of China, India, the southern part of the Soviet Union and the western United States. Probably there will always be such reservoirs—ready to be stirred up by sudden climatic change or ecological disaster. . . .

Roots of the Black Death

At least from biblical times on, there have been sporadic allusions to plagues, as well as carefully recorded outbreaks. The emperor Justinian's Constantinople, for instance, capital of the Roman empire in the East, was ravaged by plague in 541 and 542, felling perhaps 40 percent of the city's population. But none of the biblical or Roman plagues seemed so emblematic of horror and devastation as the Black Death that struck Europe in 1347. Rumors of fearful pestilence in China and throughout the East had reached Europe by 1346. "India was depopulated," reported one chronicler, "Tartary, Mesopotamia, Syria, Armenia, were covered with dead bodies; the Kurds fled in vain to the mountains. In Caramania and Caesarea none were left alive."

Untold millions would die in China and the rest of the East before the plague subsided again. By September of 1345, the *Yersinia pestis* bacillus, probably carried by rats, reached the Crimea, on the northern coast of the Black Sea, where Italian merchants had a good number of trading colonies.

From the shores of the Black Sea, the bacillus seems to have entered a number of Italian ports. The most famous account has to do with a ship that docked in the Sicilian port of Messina in

1347. According to an Italian chronicler named Gabriele de Mussis, Christian merchants from Genoa and local Muslim residents in the town of Caffa on the Black Sea got into an argument; a serious fight ensued between the merchants and a local army led by a Tatar lord. In the course of an attack on the Christians, the Tatars were stricken by plague. From sheer spitefulness, their leader loaded his catapults with dead bodies and hurled them at the Christian enemy, in hopes of spreading disease among them. Infected with the plague, the Genoese sailed back to Italy, docking first at Messina.

Although de Mussis, who never traveled to the Crimea, may be a less-than-reliable source, his underlying assumption seems sound. The plague did spread along established trade routes. (Most likely, though, the pestilence in Caffa resulted from an infected population of local rats, not from the corpses lobbed over the besieged city's walls.)

The Effect of *Y. pestis* in Humans

In any case, given enough dying rats and enough engorged and frantic fleas, it will not be long before the fleas, in their search for new hosts, leap to a human being. When a rat flea senses the presence of an alternate host, it can jump very quickly and as much as 150 times its length. The average for such jumps is about six inches horizontally and four inches straight up in the air. Once on human skin, the flea will not travel far before it begins to feed.

The first symptoms of bubonic plague often appear within several days: headache and a general feeling of weakness, followed by aches and chills in the upper leg and groin, a white coating on the tongue, rapid pulse, slurred speech, confusion, fatigue, apathy and a staggering gait. A blackish pustule usually will form at the point of the fleabite. By the third day, the lymph nodes begin to swell. Because the bite is commonly in the leg, it is the lymph nodes of the groin that swell, which is how the disease got its name. The Greek word for "groin" is *boubōn*—thus, bubonic plague. The swellings will be tender, perhaps as large as an egg. The heart begins to flutter rapidly as it tries to pump blood through swollen, suffocating tissues. Subcutaneous hemorrhaging occurs, causing purplish blotches on the skin. The victim's nervous system begins to collapse, causing dreadful pain and bizarre neurological disorders, from which the "Dance of Death" rituals that accompanied the plague may have taken their inspiration.

. . . By the fourth or fifth day, wild anxiety and terror overtake the sufferer—and then a sense of resignation, as the skin blackens and the rictus of death settles on the body.

In 1347, when the plague struck in Messina, townspeople realized that it must have come from the sick and dying crews of the ships at their dock. They turned on the sailors and drove them back out to sea—eventually to spread the plague in other ports. Messina panicked. People ran out into the fields and vineyards and neighboring villages, taking the rat fleas with them.

When the citizens of Messina, already ill or just becoming ill, reached the city of Catania, 55 miles to the south, they were at first taken in and given beds in the hospital. But as the plague began to infect Catania, the townspeople there cordoned off their town and refused—too late—to admit any outsiders. The sick, turning black, stumbling and delirious, were objects more of disgust than pity; everything about them gave off a terrible stench, it was said, their "sweat, excrement, spittle, breath, so foetid as to be overpowering; urine turbid, thick, black or red. . . ."

Panic Incited by the Black Death

Wherever the plague appeared, the suddenness of death was terrifying. Today, even with . . . the advent of AIDS, it is hard to grasp the strain that the plague put on the physical and spiritual fabric of society. People went to bed perfectly healthy and were found dead in the morning. Priests and doctors who came to minister to the sick, so the wild stories ran, would contract the plague with a single touch and die sooner than the person they had come to help. . . .

In Florence, everyone grew so frightened of the bodies stacked up in the streets that some men, called *becchini,* put themselves out for hire to fetch and carry the dead to mass graves. Having in this way stepped over the boundary into the land of the dead, and no doubt feeling doomed themselves, the *becchini* became an abandoned, brutal lot. Many roamed the streets, forcing their way into private homes and threatening to carry people away if they were not paid off in money or sexual favors.

Some people, shut up in their houses with the doors barred, would scratch a sign of the cross on the front door, sometimes with the inscription "Lord have mercy on us." In one place, two lovers were supposed to have bathed in urine every morning for protection. People hovered over latrines, breathing in the stench.

Others swallowed pus from the boils of plague victims. In Avignon, Pope Clement was said to have sat for weeks between two roaring fires. . . .

It was from a time of plague, some scholars speculate, that the nursery rhyme "Ring Around the Rosy" derives: the rose-colored "ring" being an early sign that a blotch was about to appear on the skin; "a pocket full of posies" being a device to ward off stench and (it was hoped) the attendant infection; "ashes, ashes" being a reference to "ashes to ashes, dust to dust" or perhaps to the sneezing "a-choo, a-choo" that afflicted those in

While tending to dying plague victims, priests often contracted the deadly bacteria and died shortly thereafter.

whom the infection had invaded the lungs—ending, inevitably, in "all fall down."

In Pistoia, the city council enacted nine pages of regulations to keep the plague out—no Pistoian was allowed to leave town to visit any place where the plague was raging; if a citizen did visit a plague-infested area he was not allowed back in the city; no linen or woolen goods were allowed to be imported; no corpses could be brought home from outside the city; attendance at funerals was strictly limited to immediate family. None of these regulations helped.

In Siena, dogs dragged bodies from the shallow graves and left them half-devoured in the streets. Merchants closed their shops. The wool industry was shut down. Clergymen ceased administering last rites. On June 2, 1348, all the civil courts were recessed by the city council. Because so many of the laborers had died, construction of the nave for a great cathedral came to a halt. Work was never resumed: only the smaller cathedral we know today was completed.

In Venice, it was said that 600 were dying every day. In Florence, perhaps half the population died. By the time the plague swept through, as much as one-third of Italy's population had succumbed.

In Milan, when the plague struck, all the occupants of any victim's house, whether sick or well, were walled up inside together and left to die. Such draconian measures seemed to have been partially successful—mortality rates were lower in Milan than in other cities.

The Inadequacy of Medieval Medicine

Medieval medicine was at a loss to explain all this, or to do anything about it. Although clinical observation did play some role in medical education, an extensive reliance on ancient and inadequate texts prevailed. Surgeons usually had a good deal of clinical experience but were considered mainly to be skilled craftsmen, not men of real learning, and their experience was not much incorporated into the body of medical knowledge. In 1300, Pope Boniface VIII had published a bull specifically inveighing against the mutilation of corpses. It was designed to cut down on the sale of miscellaneous bones as holy relics, but one of the effects was to discourage dissection.

Physicians, priests and others had theories about the cause of the plague. Earthquakes that released poisonous fumes, for instance. Severe changes in the Earth's temperature creating southerly winds that brought the plague. The notion that the plague was somehow the result of a corruption of the air was widely believed. It was this idea that led people to avoid foul odors by holding flowers to their noses or to try to drive out the infectious foul odors by inhaling the alternate foul odors of a latrine. Some thought that the plague came from the raining down of frogs, toads and reptiles. Some physicians believed one could catch the plague from "lust with old women."

Both the pope and the king of France sent urgent requests for help to the medical faculty at the University of Paris, then one of the most distinguished medical groups in the Western world. The faculty responded that the plague was the result of a conjunction of the planets Saturn, Mars and Jupiter at 1 P.M. on March 20, 1345, an event that caused the corruption of the surrounding atmosphere.

Ultimately, of course, most Christians believed the cause of the plague was God's wrath at sinful Man. And in those terms, to be sure, the best preventives were prayer, the wearing of crosses and participation in other religious activities. In Orvieto, the town fathers added 50 new religious observances to the municipal calendar. Even so, within five months of the appearance of the plague, Orvieto lost every second person in the town.

Preventative Measures

There was also some agreement about preventive measures one might take to avoid the wrath of God. Flight was best: away from lowlands, marshy areas, stagnant waters, southern exposures and coastal areas, toward high, dry, cool, mountainous places. It was thought wise to stay indoors all day, to stay cool and to cover any windows that admitted bright sunlight. In addition to keeping flowers nearby, one might burn such aromatic woods as juniper and ash.

The retreat to the mountains, where the density of the rat population was not as great as in urban areas, and where the weather was inimical to rats and fleas, was probably a good idea—as well as perhaps proof, of a kind, of the value of empirical observation. But any useful notion was always mixed in with such wild ideas that it got lost in a flurry of desperate (and often contrary) stratagems. One should avoid bathing because that opened the pores to attack from the corrupt atmosphere, but one should wash face and feet, and sprinkle them with rose water and vinegar. In the morning, one might eat a couple of figs with rue and filberts. One expert advised eating ten-year-old treacle mixed with several dozen items, including chopped-up snake. Rhubarb was recommended, too, along with onions, leeks and garlic. The best spices were myrrh, saffron and pepper, to be taken late in the day. Meat should be roasted, not boiled. Eggs should not be eaten hard-boiled. A certain Gentile di Foligno commended lettuce; the faculty of medicine at the University of Paris advised against it. Desserts were forbidden. One should not sleep during the day. One should sleep first on the right side, then on the left. Exercise was to be avoided because it introduced more air into the body; if one needed to move, one ought to move slowly.

By the fall of 1348, the plague began to abate. But then, just as hopes were rising that it had passed, the plague broke out again in the spring and summer of 1349 in different parts of Europe. This recurrence seemed to prove that the warm weather, and people bathing in warm weather, caused the pores of the skin to open and admit the corrupted air. In other respects, however, the plague remained inexplicable. Why did some people get it and recover, while others seemed not to have got it at all—or at least showed none of its symptoms—yet died suddenly anyway? Some people died in four or five days, others died at once. Some seemed to have contracted the plague from a friend or relative who had it, others had never been near a sick person. The sheer unpredictability of it was terrifying.

Three Different Forms of Plague

In fact, though no one would know for several centuries, there were three different forms of the plague, which ran three different courses. The first was simple bubonic plague, transmitted from rat to person by the bite of the rat flea. The second and likely most common form was pneumonic, which occurred when the bacillus invaded the lungs. After a two- or three-day incubation period, anyone with pneumonic plague would have a severe, bloody cough; the sputum cast into the air would contain *Yersinia pestis.* Transmitted through the air from person to person, pneumonic plague was fatal in 95 to 100 percent of all cases.

The third form of the plague was septocemic, and its precise etiology is not entirely understood even yet. In essence, however, it appears that in cases of septocemic plague the bacillus entered the bloodstream, perhaps at the moment of the fleabite. A rash formed and death occurred within a day, or even within hours, before any swellings appeared. Septocemic plague always turned out to be fatal. . . .

Devastation of the Pandemic

The recurrence of the plague [at least once in each decade to the end of the century] after people thought the worst was over may have been the most devastating development of all. In short, Europe was swept not only by a bacillus but also by a widespread psychic breakdown—by abject terror, panic, rage, vengefulness, cringing remorse, selfishness, hysteria, and above all, by an overwhelming sense of utter powerlessness in the face of an inescapable horror.

Recent Scientific Reassessments of the Black Death

By Norman F. Cantor

Although most historians and biologists take for granted that the Black Death was caused primarily by the bubonic plague, recent scientific reassessments of the outbreak have suggested the possibility of additional causes of the mid-fourteenth-century plague, namely anthrax. Norman F. Cantor, the author of many books on the Middle Ages and a professor of history, comparative literature, and sociology at New York University, agrees with the likelihood that anthrax was operating in addition to bubonic plague during the Black Death. In the following excerpt from his book In the Wake of the Plague: The Black Death and the World It Made, *Cantor summarizes recent scientific discoveries about the Black Death and how these findings can modify researchers' understanding of the pandemic. Citing biologists and theorists such as Graham Twigg, Edward I. Thompson, and Gunnar Karlsson, Cantor argues in support of the anthrax theory, using medical, historical, and geographic evidence to bolster this claim. Cantor goes on to highlight the scientific nature of the* Y. pestis *bacteria, surmising that, although certain strains of infectious disease continue to resist antibiotics, the chance of dying from the bubonic plague today is slim. He concludes with the possibility that those who survived the bubonic plague in the past now have ancestors who are immune to the HIV/AIDS virus.*

In spite of the incapacity of the medieval medical profession to describe securely the symptoms and course of the Black Death, historians of medicine and society have been able to determine that it involved at least bubonic plague, the same pandemic that had devastated the East Roman or Byzantine Empire

Norman F. Cantor, *In the Wake of the Plague: The Black Death and the World It Made.* New York: The Free Press, 2001.

in the sixth century A.D. and invaded the whole Mediterranean world in the third century or even earlier. The only big question on the medical side of the Black Death is whether bubonic plague was exclusively the cause of the devastation of the 1340s or whether another disease was simultaneously occurring in some parts of Europe, and particularly in England.

Reasons for Skepticism

Bubonic plague is a bacillus carried by parasites on the backs of rodents, principally but not exclusively in the Middle Ages, the species of black rat. The black rats and the plague parasites residing on them could have been disseminated by shipping in international trade. The port of Bristol was the major initial point of entry for the pestilence into England.

It is this provocative picture of these rodents scurrying inland from port cities and making long journeys through the countryside at great speed so that most of Western Europe was in pandemic conditions within a year of initial contact that raises skepticism about the conventional account of the Black Death's exclusive identification with bubonic plague. . . .

Moreover, there are peculiarities about the spread of the Black Death if it was exclusively bubonic plague that was involved. In 1984 the British zoologist Graham Twigg pointed out that the plague's impact, at least in England, was as severe in some thinly populated rural areas as in thickly settled areas. The pestilence produced almost as high a level of mortality in the winter months as in summer. These qualities do not easily conform to the view that the Black Death was exclusively bubonic plague: Parasites on the backs of rats in thinly settled areas and severe impact in cold weather are not in keeping with the common activity of fleas.

Medical historians such as Twigg also noted that mortality tales of the period around 1350 frequently described a death that occurred within three or four days of incubation, much too rapid for the much longer three-phase course of the bubonic plague. Some patients died without fever and without the buboes or welts on groins or around armpits, and to explain their deaths it was proposed, in what is still a minority opinion—although one rapidly gaining strength—that the Black Death involved or was even exclusively a rare virulent antihumanoid [without human characteristics] form of cattle disease, namely anthrax.

The Anthrax Theory

Both anthrax and bubonic plague begin with similar flulike symptoms, and the two diseases could have been conflated by contemporary doctors. And it is not hard to perceive how this anthrax-based plague—if Twigg's theory is correct—could have been spread. As Europeans cleared forests for more arable land in the thirteenth century, they did not attenuate their passion for red meat, even though the supply of wild game diminished with the forest clearing. There was an enormous increase in cattle ranching, raising of herds of beef cattle in congested conditions both on the great open ranges of northern England and the small pasturages in the southern farmlands.

Before the widespread immunizing inoculation of cattle herds in the 1950s, infectious epidemics of anthrax murrain (cattle disease) were a constant threat in cattle ranches in the transatlantic world. Modern outbreaks of infectious disease among cattle, whether rinderpest in Rhodesia in the 1890s, hoof and mouth disease in western Canada in the 1950s [and most recently in Europe in 2001], or Bovine Spongiform Encephalitis ("mad cow disease") in Britain in the 1990s, have in common an extremely rapid diffusion. What is most puzzling about the Black Death of the fourteenth century is its very rapid dissemination, a quality more characteristic of a cattle disease than a rodent-disseminated one.

That cattle were ravaged by these epidemics is certain. The question remains whether a natural anthrax mutant could be communicated to humans. The answer appears to be in the affirmative. Eating tainted meat from sick herds of cattle was a form of transmission to humans just as eating chimpanzees in what is today the Republic of Congo is believed by scientists to have started the AIDS disease in East Africa in the 1930s. The "mad cow" disease that killed about seventy in Britain in the 1990s was transmitted to humans by eating tainted meat.

But in 1995 David Herlihy rejected Twigg's thesis on the grounds there were no known outbreaks of anthrax among British cattle in the mid-fourteenth century.

Further Evidence of Anthrax

The response to Herlihy's dismissal came in 1998 from Edward I. Thompson of the University of Toronto. He cited a report in 1989 of an archeological excavation done at Soutra, seventeen miles southeast of Edinburgh, where a mass grave for Black

Death victims was located outside a medieval hospital. The excavation yielded three anthrax spores from a cesspool into which human waste was discharged.

Thompson also cited ten medieval abbeys or priories whose cattle herds were known to be diseased. To that he added evidence, drawn from a contemporary document—the smoking gun—from the decade or so before the Black Death of meat from cattle "dead of murrain" (meaning cattle disease) being sold in local markets.

Anthrax spores buried in the ground remain active for half a century or more as extremely toxic for humans. During World War II both German and Allied biomedical scientists developed anthrax to use in germ warfare. It was employed by neither side in the end, but the Allies tested their variety on an island off the Scottish coast. Fifty years after the war live spores buried in the ground there were discovered and the inhabitants of the island had to be evacuated.

To these known facts and Thompson's excellent work may be added a paper by Gunnar Karlsson of the University of Iceland. The island was hit by plague in the fourteenth century, but there appear to have been no rats in Iceland before the seventeenth century. The argument against Karlsson's no rats in Iceland thesis would run like this: Rats carrying plague-ridden fleas got off the boats from Norway or England but immediately died of the cold weather. The fleas then migrated to the nearest warm bodies, namely humans. Possible, but farfetched.

Thompson's conclusion that "bubonic plague and anthrax probably coexisted during the fourteenth century" is the best that science can currently provide.

Recent Scientific Knowledge

If medieval physicians did fail to differentiate two separate kinds of plague during the Black Death, that should not surprise us: The scientific method had not yet been invented. . . .

Today, however, we have scientific—or synchronic—means of analysis. Science has indeed told us much more about this plague of six hundred years ago than the people living then knew themselves. It has not answered every question, but it has yielded surprising specifics.

For example, a team of research doctors at the Division of Infectious Diseases, St. Joseph's Hospital and Medical Center, Paterson, New Jersey, reported in 1997:

> Plague is a zoonotic infection caused by *Yersina pestis.* . . . Animal reservoirs [carriers] include rodents, rabbits, and occasionally larger animals. Cats become ill and have spread [the] disease to man. . . . Flea bites commonly spread disease to man. Person to person spread has not been a recent feature until the purported outbreak of plague and plague pneumonia in India in 1994. Other factors that increase risk of infection in endemic areas are occupation—veterinarians and assistants, pet ownership, direct animal-reservoir contact especially during the hunting season, living in households with [a disease] case, and, mild winters, cool, moist springs, and early summers.

In 1996 a team at the Laboratory of Microbial Structure and Function, working with funding from the National Institute of Allergy and Infectious Diseases, National Health Institutes, were actually able to explain what went on within a plague-infected flea:

> *Yersinia pestis,* the cause of bubonic plague, is transmitted by the bites of infected fleas. Biological transmission of plague depends on blockage of the foregut of the flea by a mass of plague bacilli. Blockage was found to be dependent on the hermin storage (hms) locus. *Yersinia pestis* hms mutants established long-term infection of the flea's midgut but failed to colonize the proventriculus, the site in the foregut where blockage normally develops. Thus the hms locus markedly alters the course of *Y. pestis* infection in its insect vector, leading to a change in blood-feeding behavior and to efficient transmission of plague.

This is something medieval medicine did not know: the inner life of a sick flea.

Threat of the Plague Today

More important, perhaps, is that if it is identified early enough, the plague can be cured by science today. For many years I told my students at New York University that if they are taking a shower in the college gymnasium and the person in the next stall emerges with black welts under the armpits and in the groin (the infamous plague buboes) they should dress and leave immediately. And if a rat runs by as well as the buboe-marred student, don't even bother to dress. Wrap a towel around your body and head for the nearest exit. Like most things I said in class, this got a big laugh; they didn't believe me. But I was serious.

What is truly frightening, as reported in the *Royal Geographical Magazine* in 1998, is that around the world strains of infectious disease, especially tuberculosis and meningitis, not excluding bubonic plague, that are newly resistant to antibiotics are turning up. Recently a superstrength level of antibiotics has been announced to deal with the problem. It is man versus microbe in a continuing, escalating battle.

But the chance of dying of bubonic plague in the U.S.A. today is much less than the chance of being killed in an airplane crash. Don't worry—yet. And there is more good news: the Black Death may also have protected you against the current AIDS scourge.

As part of the intense study of human genetic development related to the genome project—the mapping of the total genetic structure of human beings—a team of six scientists at the National Cancer Institute's Laboratory of Genetic Diversity in 1997 made an exciting announcement. They discovered that a genetic mutant that "occurred in the order of 4,000 years ago" gave today's human carrier of this mutant (called CCR5) immunity against HIV and therefore AIDS.

One of six signatories to the accompanying article in the *American Journal of Human Genetics,* Stephen J. O'Brien, in the following year in the same journal revealed an even more startling follow-up discovery. The mutant CCR5 could in fact, said O'Brien, be traced back to only seven hundred years ago. At that time a "historic strong selective event involving a pathogen that like HIV-1 utilizes CCR5" established an immunity "in ancestral Caucasian populations." Eighteen scientists from around the world affirmed O'Brien's hypothesis.

The event he described, of course, could only be the Black Death. There is thus—if O'Brien is correct—a genetic relationship between the Black Death and AIDS. If you are descended from a Caucasian who contracted the plague of the mid-fourteenth century and that ancestor survived, you may have complete immunity to HIV/AIDS. And it is believed that up to 15 percent of the Caucasian population could fall into this lucky category.

Chapter 2

"The Whole Place Was a Sepulchre": Contemporary Chronicles of the Black Death

Death Comes to Florence

BY GIOVANNI BOCCACCIO

Giovanni Boccaccio (1313–1375) was an Italian humanist and classicist who wrote the most famous description of the Black Death in literature. The following excerpt is taken from the preface to his novel The Decameron, *a collection of stories told by a group of ten men and women who have fled the city of Florence to avoid the plague, passing their time leisurely in the countryside. Boccaccio notes that no amount of preparation or prayer could have saved the residents of Florence from the horrific buboes. Physicians, too, were useless; the plague spread so quickly that the healthy became frightened of the sick, shunning the infected, even if they were loved ones. Boccaccio describes four types of reactions to the mass devastation: There were those who lived temperately to avoid contagion, those who drank and ate with pleasure and gluttony, those who followed the suggestions of physicians and acted moderately, and those who simply fled the city in fear. No regular funerals were held, hundreds of corpses were toppled into mass trenches, and laws and work were ignored. Through Boccaccio's description of Florence, one can glean a sense of the fear and horror that must have enveloped all of Europe during the Black Death.*

I say, then, that the years of the beatific incarnation of the Son of God had reached the tale of one thousand three hundred and forty-eight, when in the illustrious city of Florence, the fairest of all the cities of Italy, there made its appearance that deadly pestilence, which, whether disesminated [spread] by the influence of the celestial bodies, or sent upon us mortals by God in His just wrath by way of retribution for our iniquities, had had its origin some years before in the East, whence, after destroying an innumerable multitude of living beings, it had propagated it-

Giovanni Boccaccio, *The Decameron*, translated by J.M. Rigg. London: The Navarre Society, 1921.

self without respite from place to place, and so, calamitously, had spread into the West.

The Form of the Malady

In Florence, despite all that human wisdom and forethought could devise to avert it, as the cleansing of the city from many impurities by officials appointed for the purpose, the refusal of entrance to all sick folk, and the adoption of many precautions for the preservation of health; despite also humble supplications addressed to God, and often repeated both in public procession and otherwise, by the devout; towards the beginning of the spring of the said year the doleful effects of the pestilence began to be horribly apparent by symptoms that shewed as if miraculous.

Not such were they as in the East, where an issue of blood from the nose was a manifest sign of inevitable death; but in men and women alike it first betrayed itself by the emergence of certain tumours in the groin or the armpits, some of which grew as large as a common apple, others as an egg, some more, some less, which the common folk called *gavoccioli.* From the two said parts of the body this deadly *gavocciolo* soon began to propagate and spread itself in all directions indifferently; after which the form of the malady began to change, black spots or livid making their appearance in many cases on the arm or the thigh or elsewhere, now few and large, now minute and numerous. And as the *gavocciolo* had been and still was an infallible token of approaching death, such also were these spots on whomsoever they shewed themselves. Which maladies seemed to set entirely at naught both the art of the physician and the virtues of physic; indeed, whether it was that the disorder was of a nature to defy such treatment, or that the physicians were at fault—besides the qualified there was now a multitude both of men and of women who practised without having received the slightest tincture of medical science—and, being in ignorance of its source, failed to apply the proper remedies; in either case, not merely were those that recovered few, but almost all within three days from the appearance of the said symptoms, sooner or later, died, and in most cases without any fever or other attendant malady.

The Nature of the Contagion

Moreover, the virulence of the pest was the greater by reason that intercourse was apt to convey it from the sick to the whole, just as fire devours things dry or greasy when they are brought close to it. Nay, the evil went yet further, for not merely by speech or association with the sick was the malady communicated to the healthy with consequent peril of common death; but any that touched the clothes of the sick or aught else that had been touched or used by them, seemed thereby to contract the disease. . . .

I say, then, that such was the energy of the contagion of the said pestilence, that it was not merely propagated from man to man, but, what is much more startling, it was frequently observed, that things which had belonged to one sick or dead of the disease, if touched by some other living creature, not of the human species, were the occasion, not merely of sickening, but of an almost instantaneous death. Whereof my own eyes (as I said a little before) had cognisance, one day among others, by the following experience. The rags of a poor man who had died of the disease being strewn about the open street, two hogs came thither, and after, as is their wont, no little trifling with their snouts, took the rags between their teeth and tossed them to and fro about their chaps; whereupon, almost immediately, they gave a few turns, and fell down dead, as if by poison, upon the rags which in an evil hour they had disturbed.

The Four Reactions

In which circumstances, not to speak of many others of a similar or even graver complexion, divers apprehensions and imaginations were engendered in the minds of such as were left alive, inclining almost all of them to the same harsh resolution, to wit, to shun and abhor all contact with the sick and all that belonged to them, thinking thereby to make each his own health secure. Among whom there were those who thought that to live temperately and avoid all excess would count for much as a preservative against seizures of this kind. Wherefore they banded together, and, dissociating themselves from all others, formed communities in houses where there were no sick, and lived a separate and secluded life, which they regulated with the utmost care, avoiding every kind of luxury, but eating and drinking very moderately of the most delicate viands and the finest wines, holding converse with none but one another, lest tidings of sickness or death should reach them, and diverting their minds with music and such other delights as they could devise. Others, the bias of whose minds was in the opposite direction, maintained, that to drink freely, frequent places of public resort, and take their pleasure with song and revel, sparing to satisfy no appetite, and to laugh and mock at no event, was the sovereign remedy for so great an evil: and that which they affirmed they also put in practice, so far as they were able, resorting day and night, now to this tavern, now to that, drinking with an entire disregard of rule or measure, and by preference making the houses of others, as it were, their inns. . . . Thus, adhering ever to their inhuman determination to shun the sick, as far as possible, they ordered their life. In this extremity of our city's suffering and tribulation the venerable authority of laws, human and divine, was abased and all but totally dissolved, for lack of those who should have administered and enforced them, most of whom, like the rest of the citizens, were either dead or sick, or so hard bested for servants that they were unable to execute any office; whereby every man was free to do what was right in his own eyes.

Not a few there were who belonged to neither of the two said parties, but kept a middle course between them, neither laying the same restraint upon their diet as the former, nor allowing themselves the same license in drinking and other dissipations as the latter, but living with a degree of freedom sufficient to satisfy their appetites, and not as recluses. They therefore walked abroad, carrying in their hands flowers or fragrant herbs or divers sorts of spices, which they frequently raised to their noses, deeming it an excellent thing thus to comfort the brain with such perfumes, because the air seemed to be everywhere laden and reeking with the stench emitted by the dead and the dying, and the odours of drugs.

Some again, the most sound, perhaps, in judgment, as they were also the most harsh in temper, of all, affirmed that there was no medicine for the disease superior or equal in efficacy to flight; following which prescription a multitude of men and women, negligent of all but themselves, deserted their city, their houses, their estates, their kinsfolk, their goods, and went into voluntary exile, or migrated to the country parts, as if God in visiting men with this pestilence in requital of their iniquities would not pursue them with His wrath wherever they might be, but intended the destruction of such alone as remained within the circuit of the walls of the city; or deeming, perchance, that it was now time for all to flee from it, and that its last hour was come.

Abandoning the Sick

Of the adherents of these divers opinions not all died, neither did all escape; but rather there were, of each sort and in every place, many that sickened, and by those who retained their health were treated after the example which they themselves, while whole, had set, being everywhere left to languish in almost total neglect. Tedious were it to recount, how citizen avoided citizen, how among neighbours was scarce found any that shewed fellow-feeling for another, how kinsfolk held aloof, and never met, or but rarely; enough that this sore affliction entered so deep into the minds of men and women, that in the horror thereof brother was forsaken by brother, nephew by uncle, brother by sister, and oftentimes husband by wife; nay, what is more, and scarcely to be believed, fathers and mothers were found to abandon their own children, untended, unvisited, to their fate, as if they had been strangers. Wherefore the sick of both sexes, whose number could not be estimated, were left without resource but in the charity of friends (and few such there were), or the interest of servants.... In consequence of which dearth of servants and dereliction of the sick by neighbours, kinsfolk and friends, it came to pass—a thing, perhaps, never before heard of—that no woman, however dainty, fair or well-born she might be, shrank, when stricken with the disease, from the ministrations of a man, no matter whether he were young or no, or scrupled to expose to him every part of her body, with no more shame than if he had been a woman, submitting of necessity to that which her malady required; wherefrom, perchance, there resulted in after time some loss of modesty in such as recovered. Besides which many succumbed, who with proper attendance, would, perhaps, have escaped death; so that, what with the virulence of the plague and the lack of due tendance of the sick, the multitude of the deaths, that daily and nightly took place in the city, was such that those who heard the tale—not to say witnessed the fact—were struck dumb with amazement. Whereby, practices contrary to the former habits of the citizens could hardly fail to grow up among the survivors.

A New Brand of Funeral

It had been, as today it still is, the custom for the women that were neighbours and of kin to the deceased to gather in his house with the women that were most closely connected with him, to wail with them in common, while on the other hand his male kinsfolk and neighbours, with not a few of the other citizens, and a due proportion of the clergy according to his quality, assembled without, in front of the house, to receive the corpse; and so the dead man was borne on the shoulders of his peers, with funeral pomp of taper and dirge, to the church selected by him before his death. Which rites, as the pestilence waxed in fury, were either in whole or in great part disused, and gave way to others of a novel order. For not only did no crowd of women surround the bed of the dying, but many passed from this life unregarded, and few indeed were they to whom were accorded the lamentations and bitter tears of sorrowing relations; nay, for the most part, their place was taken by the laugh, the jest, the festal gathering; observances which the women, domestic piety in large measure set aside, had adopted with very great advantage to their health. Few also there were whose bodies were attended to the church by more than ten or twelve of their neighbours, and those not the honourable and respected citizens; but a sort of corpse-carriers drawn from the baser ranks, who called themselves *becchini* and performed such offices for hire, would shoulder the bier, and with hurried steps carry it, not to the church of the dead man's choice, but to that which was nearest at hand, with four or six priests in front and a candle or two, or, perhaps, none; nor did the priests distress themselves with too long and solemn an office, but with the aid of the *becchini* hastily consigned the corpse to the first tomb which they found untenanted. The condition of the lower, and, perhaps, in great measure of the middle ranks, of the people shewed even worse and more deplorable; for, deluded by hope or constrained by poverty, they stayed in their quarters, in their houses, where they sickened by thousands a day, and, being without service or help of any kind, were, so to speak, irredeemably devoted to the death which overtook them. Many died daily or nightly in the public streets; of many others, who died at home, the departure was hardly observed by their neighbours, until the stench of their putrefying bodies carried the tidings; and what with their corpses and the corpses of others who died on every hand the whole place was a sepulchre.

Burying the Dead

It was the common practice of most of the neighbours, moved no less by fear of contamination by the putrefying bodies than by charity towards the deceased, to drag the corpses out of the houses with their own hands, aided, perhaps, by a porter, if a porter was to be had, and to lay them in front of the doors, where any one who made the round might have seen, especially in the morning, more of them than he could count; afterwards they would have biers brought up, or, in default, planks, whereon they laid them. Nor was it once or twice only that one and the same bier carried two or three corpses at once; but quite a considerable number of such cases occurred, one bier sufficing for husband and wife, two or three brothers, father and son, and so forth. . . . Nor, for all their number, were their obsequies honoured by either tears or lights or crowds of mourners; rather, it was come to this, that a dead man was then of no more account than a dead goat would be today. . . .

As consecrated ground there was not in extent sufficient to provide tombs for the vast multitude of corpses which day and night, and almost every hour, were brought in eager haste to the churches for interment, least of all, if ancient custom were to be observed and a separate resting-place assigned to each, they dug, for each graveyard, as soon as it was full, a huge trench, in which they laid the corpses as they arrived by hundreds at a time, piling them up as merchandise is stowed in the hold of a ship, tier upon tier, each covered with a little earth, until the trench would hold no more.

Misery Everywhere

But I spare to rehearse with minute particularity each of the woes that came upon our city, and say in brief, that, harsh as was the tenor of her fortunes, the surrounding country knew no mitigation; for there—not to speak of the castles, each, as it were, a little city in itself—in sequestered village, or on the open champaign, by the wayside, on the farm, in the homestead, the poor hapless husbandmen and their families, forlorn of physicians' care or servants' tendance, perished day and night alike, not as men, but rather as beasts. Wherefore, they too, like the citizens, abandoned all rule of life, all habit of industry, all counsel of prudence; nay, one and all, as if expecting each day to be their last, not merely ceased to aid Nature to yield her fruit in due season of their beasts and their lands and their past labours, but left no means unused, which ingenuity could devise, to waste their accumulated store; denying shelter to their oxen, asses, sheep, goats, pigs, fowls, nay, even to their dogs, man's most faithful companions, and driving them out into the fields to roam at large amid the unsheaved, nay, unreaped corn. . . . But enough of the country! What need we add, but (reverting to the city) that such and so grievous was the harshness of heaven, and perhaps in some degree of man, that, what with the fury of the pestilence, the panic of those whom it spared, and their consequent neglect or desertion of not a few of the stricken in their need, it is believed without any manner of doubt, that between March and the ensuing July upwards of a hundred thousand human beings lost their lives within the walls of the city of Florence, which before the deadly visitation would not have been supposed to contain so many people! How many grand palaces, how many stately homes, how many splendid residences, once full of retainers, of lords, of ladies, were now left desolate of all, even to the meanest servant! How many families of historic fame, of vast ancestral domains, and wealth proverbial, found now no scion to continue the succession! How many brave men, how many fair ladies, how many gallant youths, whom any physician, were he [famous Greek physicians and theoreticians] Galen, Hippocrates, or Æsculapius himself, would have pronounced in the soundest of health, broke fast with their kinsfolk, comrades and friends in the morning, and when evening came, supped with their forefathers in the other world!

The Impact of the Black Death in England

By Henry Knighton

Henry Knighton worked as a clergyman at St. Mary's Abbey in Leicester at the time of the Black Death. Like many other contemporary chroniclers, Knighton links earthquakes with the epidemic in an attempt to explain the disaster in rational terms. Also like other chroniclers, Knighton describes the deaths of many priests during the Black Death. In response, many widowers rushed to fill the vacancies in the clergy, but these laymen's lack of education and preparation was not well received. Knighton's account of the impact of the plague in England is unique for a few reasons. He cites the mortality of sheep all across England—a statistic inconsistent with the nature of bubonic plague and possibly in support of Graham Twigg's theory that anthrax was operating in addition to bubonic plague at the time. Knighton's account is important, moreover, for its attention to the economic consequences wrought by the plague; the price of cattle became excessively cheap, while the price of hired labor became increasingly expensive. The aristocracy took desperate measures to control this social and economic chaos. According to Knighton, the king punished those serfs who charged exorbitant prices for their services, and landlords decreased their rents in the hopes that tenants would stay.

In this year [1348] and in the following one there was a general mortality of men throughout the whole world. It first began in India, then in Tharsis [Taurus?], then it came to the Saracens, [i.e., Muslims] and finally to the Christians and Jews, so that in the space of one year, from Easter to Easter, as the rumour spread in the Roman curia, there had died, as if by sudden death,

Henry Knighton, "The Impact of the Black Death," *The Portable Medieval Reader*, edited by James B. Ross and Mary M. McLaughlin. New York: The Viking Press, 1949.

in those remote regions eight thousand legions, besides the Christians. The king of Tharsis, seeing such a sudden and unheard-of slaughter of his people, began a journey to Avignon with a great multitude of his nobles, to propose to the pope that he would become a Christian and be baptized by him, thinking that he might thus mitigate the vengeance of God upon his people because of their wicked unbelief. Then, when he had journeyed for twenty days, he heard that the pestilence had struck among the Christians, just as among other peoples. So, turning in his tracks, he travelled no farther but hastened to return home. The Christians, pursuing these people from behind, slew about seven thousand of them.

The Epidemic Linked to Earthquakes?

There died in Avignon in one day one thousand three hundred and twelve persons, according to a count made for the pope, and, another day, four hundred persons and more. Three hundred and fifty-eight of the Friars Preachers in the region of Provence died during Lent. At Montpellier, there remained out of a hundred and forty friars only seven. There were left at Magdalena only seven friars out of a hundred and sixty, and yet enough. At Marseilles, of a hundred and fifty Friars Minor, there remained only one who could tell the others; that was well, indeed. Of the Carmelites, more than a hundred and sixty-six had died at Avignon before the citizens found out what had happened. For they believed that one had killed another. There was not one of the English Hermits left in Avignon. . . .

At this same time the pestilence became prevalent in England, beginning in the autumn in certain places. It spread throughout the land, ending in the same season of the following year. At the same time many cities in Corinth and Achaia were overturned, and the earth swallowed them. Castles and fortresses were broken, laid low, and swallowed up. Mountains in Cyprus were levelled into one, so that the flow of the rivers was impeded, and many cities were submerged and villages destroyed. Similarly, when a certain friar was preaching at Naples, the whole city was destroyed by an earthquake. Suddenly, the earth was opened up, as if a stone had been thrown into water, and everyone died along with the preaching friar, except for one friar who, fleeing, escaped into a garden outside the city. All of these things were done by an earthquake. . . .

Social and Economic Chaos

Then that most grievous pestilence penetrated the coastal regions [of England] by way of Southampton, and came to Bristol, and people died as if the whole strength of the city were seized by sudden death. For there were few who lay in their beds more than three days or two and a half days; then that savage death snatched them about the second day. In Leicester, in the little parish of St. Leonard, more than three hundred and eighty died; in the parish of the Holy Cross, more than four hundred, and in the parish of St. Margaret in Leicester, more than seven hundred. And so in each parish, they died in great numbers. Then the bishop of Lincoln sent through the whole diocese, and gave the general power to each and every priest, both regular and secular, to hear confessions and to absolve, by the full and entire power of the bishop, except only in the case of debt. And they might absolve in that case if satisfaction could be made by the person while he lived, or from his property after his death. Likewise, the pope granted full remission of all sins, to be absolved completely, to anyone who was in danger of death, and he granted this power to last until the following Easter. And everyone was allowed to choose his confessor as he pleased.

During this same year, there was a great mortality of sheep everywhere in the kingdom; in one place and in one pasture, more than five thousand sheep died and became so putrefied that neither beast nor bird wanted to touch them. And the price of everything was cheap, because of the fear of death; there were very few who took any care for their wealth, or for anything else. For a man could buy a horse for half a mark, which before was worth forty shillings, a large fat ox for four shillings, a cow for twelve pence, a heifer for sixpence, a large fat sheep for four pence, a sheep for threepence, a lamb for two pence, a fat pig for five pence, a stone of wool for nine pence. And the sheep and cattle wandered about through the fields and among the crops, and there was no one to go after them or to collect them. They perished in countless numbers everywhere, in secluded ditches and hedges, for lack of watching, since there was such a lack of serfs and servants, that no one knew what he should do. For there is no memory of a mortality so severe and so savage from the time of Vortigern, king of the Britons, in whose time, as [early medieval chronicler] Bede says, the living did not suffice to bury the dead. In the following autumn, one could not hire a reaper at a lower wage than eight pence with food, or a mower at less than twelve pence with food. Because of this, much grain rotted in the fields for lack of harvesting, but in the year of the plague, as was said above, among other things there was so great an abundance of all kinds of grain that no one seemed to have concerned himself about it.

Crisis in the Church

The Scots, hearing of the cruel pestilence in England, suspected that this had come upon the English by the avenging hand of God, and when they wished to swear an oath, they swore this one, as the vulgar rumour reached the ears of the English, "be the foul deth of Engelond." And so the Scots, believing that the horrible vengeance of God had fallen on the English, came together in the forest of Selkirk to plan an invasion of the whole kingdom of England. But savage mortality supervened, and the sudden and frightful cruelty of death struck the Scots. In a short time, about five thousand died; the rest, indeed, both sick and well, prepared to return home, but the English, pursuing them, caught up with them, and slew a great many of them.

Master Thomas Bradwardine was consecrated archbishop of Canterbury by the pope, and when he returned to England, came to London. In less than two days he was dead. He was famous above all other clerks in Christendom, in theology especially, but also in other liberal studies. At this same time there was so great a lack of priests everywhere that many widowed churches had no divine services, no masses, matins, vespers, sacraments, and sacramentals. One could hardly hire a chaplain to minister to any church for less than ten pounds or ten marks, and whereas, before the pestilence, when there were plenty of priests, one could hire a chaplain for five or four marks or for two marks, with board, there was scarcely anyone at this time who wanted to accept a position for twenty pounds or twenty marks. But within a short time a very great multitude whose wives had died of the plague rushed into holy orders. Of these many were illiterate and, it seemed, simply laymen who knew nothing except how to read to some extent. The hides of cattle went up from a low price to twelve pence, and for shoes the price went to ten, twelve, fourteen pence; for a pair of leggings, to three and four shillings.

Meanwhile, the king ordered that in every county of the kingdom, reapers and other labourers should not receive more than

Terrified of exposing themselves to the Black Death, people often abandoned the bodies of dead plague victims in the streets.

they were accustomed to receive, under the penalty provided in the statute, and he renewed the statute from this time. The labourers, however, were so arrogant and hostile that they did not heed the king's command, but if anyone wished to hire them, he had to pay them what they wanted, and either lose his fruits and crops or satisfy the arrogant and greedy desire of the labourers as they wished. When it was made known to the king that they had not obeyed his mandate, and had paid higher wages to the labourers, he imposed heavy fines on the abbots, the priors, the great lords and the lesser ones, and on others both greater and lesser in the kingdom. From certain ones he took a hundred shillings, from some, forty shillings, from others, twenty shillings, and from each according to what he could pay. And he took from each plough-land in the whole kingdom twenty shillings, and not one-fifteenth less than this. Then the king had many labourers arrested, and put them in prison. Many such hid themselves and ran away to the forests and woods for a while, and those who were captured were heavily fined. And the greater number swore that they would not take daily wages above those set by ancient custom, and so they were freed from prison. It was done in like manner concerning other artisans in towns and villages. . . .

After the aforesaid pestilence, many buildings, both large and small, in all cities, towns, and villages had collapsed, and had completely fallen to the ground in the absence of inhabitants. Likewise, many small villages and hamlets were completely deserted; there was not one house left in them, but all those who had lived in them were dead. It is likely that many such hamlets will never again be inhabited. In the following summer [1350], there was so great a lack of servants to do anything that, as one believed, there had hardly been so great a dearth in past times. For all the beasts and cattle that a man possessed wandered about without a shepherd, and everything a man had was without a caretaker. And so all necessities became so dear that anything that in the past had been worth a penny was now worth four or five pence. Moreover, both the magnates of the kingdom and the other lesser lords who had tenants, remitted something from the rents, lest the tenants should leave, because of the lack of servants and the dearth of things. Some remitted half the rent, some more and others less, some remitted it for two years, some for three, and others for one year, according as they were able to come to an agreement with their tenants. Similarly, those who received day-work from their tenants throughout the year, as is usual from serfs, had to release them and to remit such services. They either had to excuse them entirely or had to fix them in a laxer manner at a small rent, lest very great and irreparable damage be done to the buildings, and the land everywhere remain completely uncultivated. And all foodstuffs and all necessities became exceedingly dear.

A Treatise on the Prevention and Cure of the Plague

BY JOHN OF BURGUNDY

In an attempt to explain the causes of the plague, as well as to offer suggestions for preventative and curative measures, many physicians at the time of the Black Death wrote "tractates"—treatises about their medical theories for the benefit of the common people. The following excerpt from John of Burgundy's 1365 Treatise on the Epidemic Sickness *is one of the most popular of them all. John of Burgundy begins by stressing the importance of astrology in understanding medicine, and he goes on to suggest ways to avoid contracting the plague. He believes that one must avoid baths, overindulgence in food and drink, and sexual intercourse, but he encourages the lighting of fires in order to inhale herbal concoctions. Because many medieval physicians believed that the plague was caused by corrupted air that had poisoned the bloodstream, John of Burgundy suggests immediate bloodletting (or the draining of blood) as a remedy for those infected with the plague. His directions for bloodletting are detailed; one must drain veins only on the side of the body that contains the swellings and buboes, and one must drain blood almost immediately after contagion, otherwise the process is ineffective. John of Burgundy ends his treatise with an appreciation of the progress made in medicine since the time of Hippocrates, the Greek Father of Medicine.*

Everything below the moon, the elements and the things compounded of the elements, is ruled by the things above, and the highest bodies are believed to give being, nature, substance, growth and death to everything below their spheres. It was, therefore, by the influence of the heavenly bod-

John of Burgundy, "The Treatise of John of Burgundy, 1365," *The Black Death*, edited and translated by Rosemary Horrox. Manchester, UK: Manchester University Press, 1994.

ies that the air was recently corrupted and made pestilential. . . . The result was a widespread epidemic, traces of which still remain in several places. Many people have been killed, especially those stuffed full of evil humours, for the cause of the mortality is not only the corruption of the air, but the abundance of corrupt humours within those who die of the disease. For as [Greek physician and writer] Galen says in the book of fevers, the body suffers no corruption unless the material of the body has a tendency towards it, and is in some way subject to the corruptive cause; for just as fire only takes hold on combustible material, so pestilential air does no harm to a body unless it finds a blemish where corruption can take hold. . . .

The Importance of Astrology

Hippocrates [the Greek physician known as the Father of Medicine] testifies in his *Epidemia* . . . that no one ought to be put under the care of any physician who is ignorant of astrology. For the arts of medicine and astrology balance each other, and in many respects one science supports the other in that one cannot be understood without the other. I am as a result convinced by practical experience that medicine—however well it has been compounded and chosen according to medical rules—does not work as the practitioner intends and is of no benefit to the patient if it is given when the planets are contrary. Thus if medicine is given as a laxative it should be with reference to the planets if the patient is to empty his bowels successfully, and also if he is not to have an adverse reaction to the medicine. Accordingly those who have drunk too little of the nectar of astrology cannot offer a remedy for epidemic diseases. Because they are ignorant of the cause and quality of the disease they cannot cure it; for as the prince of doctors says: 'How can you cure, if you are ignorant of the cause?'. . .

It is accordingly obvious that physic is of little effect without astrology, and as a result of a lack of advice many succumb to disease. And therefore I, John of Burgundy, otherwise known as Bearded John, citizen of Liège and practitioner of the art of medicine, although the least of physicians, produced a treatise at the beginning of this epidemic on the causes and nature of corrupt air, of which many people acquired copies. I also published a treatise on the difference between epidemic and other illness. Anyone who has copies will find many things in these treatises about lifestyle and cures—but not everything about cures. Because the epidemic is now newly returned, and will return again in future because it has not yet run its course, and because I pity the carnage among mankind and support the common good and desire the health of all, and have been moved by a wish to help, I intend, with God's help, to set out more clearly in this schedule the prevention and cure of these illnesses, so that hardly anyone should have to resort to a physician but even simple folk can be their own physician, preserver, ruler and guide.

Advice for Prevention

First, you should avoid over-indulgence in food and drink, and also avoid baths and everything which might rarefy the body and open the pores, for the pores are the doorways through which poisonous air can enter, piercing the heart and corrupting the life force. Above all sexual intercourse should be avoided. You should eat little or no fruit, unless it is sour, and should consume easily-digested food and spiced wine diluted with water. Avoid mead and everything else made with honey, and season food with strong vinegar. In cold or rainy weather you should light fires in your chamber and in foggy or windy weather you should inhale aromatics every morning before leaving home: ambergris, musk, rosemary and similar things if you are rich; zedoary, cloves, nutmeg, mace and similar things if you are poor. . . . Later, on going to bed, shut the windows and burn juniper branches, so that the smoke and scent fills the room. Or put four live coals in an earthenware vessel and sprinkle a little of the following powder on them and inhale the smoke through mouth and nostrils before going to sleep: take white frankincense, labdanum, storax, calaminth, and wood of aloes and grind them to a very fine powder. And do this as often as a foetid or bad odour can be detected in the air, and especially when the weather is foggy or the air tainted, and it can protect against the epidemic.

If, however, the epidemic occurs during hot weather it becomes necessary to adopt another regimen, and to eat cold things rather than hot and also to eat more sparingly than in cold weather. You should drink more than you eat, and take white wine with water. You should also use large amounts of vinegar and verjuice in preparing food, but be sparing with hot substances such as pepper, galingale or grains of paradise. Before leaving home in the morning smell roses, violets, lilies, white and red sandalwood, musk or camphor if the weather is misty or the air quality bad. . . .

If you should feel a motion of the blood like a fluttering or prickling, let blood from the nearest vein on the same side of the body, and the floor of the room in which you are lying should be sprinkled two or three times a day with cold water and vinegar, or with rose water if you can afford it. . . .

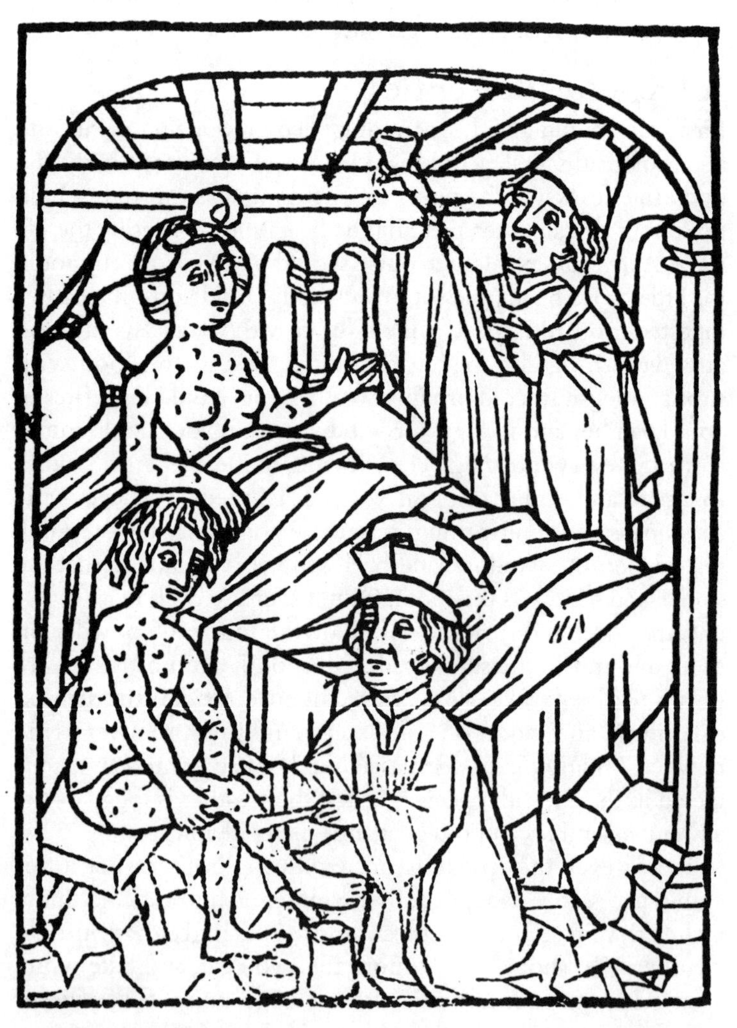

As depicted in this archaic woodcut, many medieval physicians treated plague victims with the practice of bloodletting, which was believed to draw out the poisons from the bloodstream.

Bloodletting as Remedy

Now if anyone should contract epidemic disease for lack of a good regimen it is necessary to look at remedies and at how he should proceed, for these epidemic diseases take hold in twenty four hours and it is therefore vital to apply a remedy immediately. But first it should be understood that there are three principal members in the human body: the heart, the liver and the brain, and that each of these has its emunctory, where it expels its waste matter. Thus the armpits are the emunctories of the heart, the groin for the liver, and under the ears or beneath the tongue for the brain. Now it is necessary to know that it is the nature of poison to descend from the stomach, as is shown by the bite of a serpent or other venomous creature. And thus poisonous air, when it has been mixed with blood, immediately seeks the heart, the seat of nature, to attack it. The heart, sensing the injury, labours to defend itself, driving the poisoned blood to its emunctory. If then the venomous matter finds its way blocked, so that it cannot ascend back to the heart by some other path, it seeks another principal member, the liver, so that it can destroy that. The liver, fighting back, drives the resinous matter to its emunctory. In the same way it lays claim to the brain. By means of these events, which are signs to the physician, it is possible to tell where the poisonous matter is lurking and by what vein it ought to be drained.

For if the infected blood is driven to the armpits it can be deduced that the heart is oppressed and suffering, and so blood should be let immediately from the cardiac vein, but on the same side of the body, not the opposite side, for that would do double damage: firstly, the good and pure blood on the uncorrupted side would be drained away; secondly, the corrupt and poisoned blood would be thereby drawn to the healthy side of the body, with the result that the blood on both sides would become corrupted. What is worse, in the process the venomous blood would pass through the region of the heart and infect it, and thereby cause the rapid onset of illness. If, however, the patient feels prickings in the region of the liver, blood should be let immediately from the basilic vein of the right arm (that is the vein belonging to the liver, which is immediately below the vein belonging to the heart) or in the cephalic vein of the right hand, which is between the third and little fingers.

If the liver expels matter to the groin, and it becomes visible next to the privy member towards the inside of the leg, then a vein should be opened in the foot on the same side of the body, between the big toe and the toe next to it, for if blood was let in the arm it would again draw the matter to the heart, which would be a major error. If the poison manifests itself more towards the flank, and further from the genitals, then open a vein in the foot on the same side of the body between the little toe and the toe next to it, or the vein next to the ankle or heel of that foot, or scarify [cauterize] the leg next to the tumour with pitch [medicinal resin]. If the poison appears at the emunctories of the brain, let blood from the cephalic vein above the median vein in the arm on the same side of the body, or from the vein in the hand between the thumb and index finger, or scarify the flesh with pitch between the shoulder blades. . . .

The Importance of Bloodletting

I say that these pestilential illnesses have a short and sudden beginning and a rapid development, and therefore in these illnesses those who wish to work a cure ought not to delay, and bleeding, which is the beginning of the cure, should not be put off until the first or second day. On the contrary, if someone can be found to do it, blood should be taken from the vein going from the seat of the diseased matter (that is, in the place where morbidity has appeared) in the very hour in which the patient was seized by illness. And if the bleeding cannot be done within the hour, at least let it be done within six hours, and if that is not possible then do not let the patient eat or drink until the bleeding has been done. But do not by any means delay the bleeding for longer than twelve hours, for if it is done within twelve hours, while the poisonous matter is still moving about the body, it will certainly save the patient. But if it is delayed until the illness is established, and then done, it will certainly do no harm but there is no certainty that it will rescue the patient from danger, for by then the bad blood will be so clotted and thickened that it will be scarcely able to flow from the vein. If, after the phlebotomy, the poisonous matter spreads again, the bleeding should be repeated in the same vein or in another going from the seat of the diseased matter. Afterwards three or five spoonsful of . . . herbal water . . . should be administered. . . .

And many people have been cured by one bleeding alone, performed at the right time, without any other medicine. But where people delay bleeding beyond the development of the disease, it is doubtful whether it will lead to a cure or not. For while nature keeps the matter in motion, and the heart by its expulsive virtue drives the noisome and infected blood to its emunctory, phlebotomy should be performed because it helps nature, in that the extraction and evacuation of blood strengthens the expulsive virtue of the heart and diminishes the quantity of unhealthy matter, whereby nature is made more powerful against what remains and medicine becomes more efficacious.

The Progress of Medicine

I have never known anyone treated with this type of bleeding who has not escaped death, provided that he has looked after himself well and has received substances to strengthen his heart. As a result I make bold to say—not in criticism of past authorities, but out of long experience in the matter—that modern masters are more experienced in treating pestilential epidemic diseases than all the doctors and medical experts from Hippocrates downward. For none of them saw an epidemic reigning in their time, apart from Hippocrates in the city of *Craton* and that was short-lived. Nevertheless, he drew on what he had seen in his book on epidemics. However, [famous physicians] Galen, Dioscorides, Rhazes 'Damascenus', Geber, Mesue, Copho, Constantine, Serapion, Avicenna, Algazel and all their successors never saw a general or long-lasting epidemic, or tested their cures by long experience; although they draw on the sayings of Hippocrates to discuss many things concerning epidemics. As a result, the masters of the present day are more practised in these diseases than their predecessors, for it is said, and with truth, that experience makes skill. Moved by piety and by pity for the destruction of men, I have accordingly compiled this compendium and have specified and set out the veins to bleed in these epidemic diseases, so that anyone may be his own physician. And because these illness run their course very quickly, and the poisonous matter rages through the body, let the bleeding be done without delay according to my advice, for in many cases delay brings danger.

I have composed and compiled this work not for money but for prayers, and so let anyone who has recovered from the disease pray strongly for me to our Lord God, to whom be the praise and glory throughout the whole world for ever and ever, amen.

Here ends the valuable treatise of Master John of Burgundy against epidemic disease.

Chapter 3

“The World Turned Upside Down”: Reactions to and Repercussions of the Black Death

Desperate Responses: The Flagellants and the Persecution of Jews

By Philip Ziegler

Philip Ziegler was born in 1929 and served as editorial director at William Collins for more than fifteen years. In the following excerpt from his seminal book called The Black Death, *published in 1969, Ziegler discusses two immediate and extreme reactions to the catastrophe of the Black Death that originated in Germany. The Flagellants were a group of Christian extremists who moved from town to town performing self-castigating rituals. Believing that the plague was a result of God's wrath, they scourged their own flesh with whips and spikes in the hopes that self-punishment would satiate God's anger and end the pestilence. Because they acted without the approval of the Church and claimed supernatural powers, Pope Clement VI issued a papal bull condemning flagellation. The Brethren of the Cross, as they were called, eventually disappeared. Another extremist reaction to the Black Death was the persecution of Jewish communities all across Europe. Looking for a scapegoat on which to blame their suffering, many Christians invented far-fetched claims that the Jews were poisoning their wells with the deadly disease. They tortured many Jews into confessing to the crime and massacred whole, helpless communities. Religious and governmental authorities attempted to end the persecution, but to little avail. Ziegler believes that the Flagellant movement and the killing of Jews are two examples of the man-made destruction that can occur when a culture is faced with such inexplicable tragedy.*

The Flagellant Movement, even though it dislocated life over a great area of Europe and at one time threatened the security of governments, did not, in the long run, amount to very much. It might reasonably be argued that, in a book covering so immense a subject as the Black Death, it does not merit considered attention. In statistical terms this might be true. But the Flagellants, with their visions and their superstitions, their debauches and their discipline, their idealism and their brutality, provide a uniquely revealing insight into the mind of medieval man when confronted with overwhelming and inexplicable catastrophe. Only a minority of Europeans reacted with the violence of the Flagellants but the impulses which drove this minority on were everywhere at work. To the more sophisticated the excesses of the Flagellants may have seemed distasteful; to the more prudent, dangerous. But to no one did they seem meaningless or irrelevant—that there was method in their madness was taken for granted even by the least enthusiastic. It is this, the fact that some element of the Flagellant lurked in the mind of every medieval man, which, more than the movement's curious nature and intrinsic drama, justifies its consideration in some detail. . . .

Origins of the Flagellant Movement

The practice of self-scourging as a means of mortifying the flesh seems to be first recorded in Europe in certain Italian monastic communities early in the eleventh century. As a group activity it was not known for another two hundred years. At this point, in the middle of the thirteenth century, a series of disasters convinced the Italians that God's anger had been called down on man as a punishment for his sins. The idea that he might be placated if a group of the godly drew together to protest their penitence and prove it by their deeds seems first to have occurred to a Perugian hermit called Raniero. The project was evidently judged successful, at any rate sufficiently so for the experiment to be repeated in 1334 and again a few years later. . . .

The pilgrimage of 1260 drew its authority from a Heavenly Letter brought to earth by an angel which stated that God, incensed by man's failure to observe the Sabbath day, had scourged Christendom and would have destroyed the world altogether but for the intercession of the angels and the Virgin and the altogether becoming behaviour of the Flagellants. Divine grace would be forthcoming for all those who became members of the

Brotherhood: Anybody else, it was clear, was in imminent danger of hellfire. A second edition of this letter was issued in time for the Black Death by an angel who was said to have delivered it in the Church of St Peter in Jerusalem some time in 1343. The text was identical with the first except for an extra paragraph specifically pointing out that the plague was the direct punishment of God and that the aim of the Flagellants was to induce God to relent.

The 'Brotherhood of the Flagellants' or 'Brethren of the Cross' as the movement was called in 1348, traditionally originated in Eastern Europe. . . . It was in Germany that the Flagellant Movement really took root. It is hard to be sure whether this was the result of circumstances or of the nature of the inhabitants. . . . The actual mechanism of recruitment to the Brotherhood is still obscure but the appearance of the Flagellants on the march is well attested. They moved in a long crocodile, two-by-two, usually in groups of two or three hundred but occasionally even more than a thousand strong. Men and women were segregated, the women taking their place towards the rear of the procession. At the head marched the group Master and two lieutenants carrying banners of purple velvet and cloth of gold. Except for occasional hymns the marchers were silent, their heads and faces hidden in cowls, their eyes fixed on the ground. They were dressed in sombre clothes with red crosses on back, front and cap.

The Nature of the Ritual

Word would travel ahead and, at the news that the Brethren of the Cross were on the way, the bells of the churches would be set ringing and the townsfolk pour out to welcome them. The first move was to the church where they would chant their special litany. A few parish priests used to join in and try to share the limelight with the invaders, most of them discreetly lay low until the Flagellants were on the move again. Only a handful were so high-principled or fool hardy as to deny the use of their church for the ceremony and these were usually given short shrift by the Brethren and by their own parishioners.

Sometimes the Flagellants would use the church for their own rites as well as for the litany but, provided there was a market place or other suitable site, they preferred to conduct their service in the open air. Here the real business of the day took place. A large circle was formed and the worshippers stripped to the waist, retaining only a linen cloth or skirt which stretched as far as their ankles. Their outer garments were piled up inside the circle and the sick of the village would congregate there in the hope of acquiring a little vicarious merit. On one occasion, at least, a dead child was laid within the magic circle—presumably in the hope of regeneration. The Flagellants marched around the circle; then, at a signal from the Master, threw themselves to the ground. The usual posture was that of one crucified but those with especial sins on their conscience adopted appropriate attitudes: an adulterer with his face to the ground, a perjurer on one side holding up three fingers. The Master moved among the recumbent bodies, thrashing those who had committed such crimes or who had offended in some way against the discipline of the Brotherhood.

Then came the collective flagellation. Each Brother carried a heavy scourge with three or four leather thongs, the thongs tipped with metal studs. With these they began rhythmically to beat their backs and breasts. Three of the Brethren acting as cheerleaders, led the ceremonies from the centre of the circle while the Master walked among his flock, urging them to pray to God to have mercy on all sinners. Meanwhile the worshippers kept up the tempo and their spirits by chanting the Hymn of the Flagellants. The pace grew. The Brethren threw themselves to the ground, then rose again to continue the punishment; threw themselves to the ground a second time and rose for a final orgy of self-scourging. Each man tried to outdo his neighbour in pious suffering, literally whipping himself into a frenzy in which pain had no reality. Around them the townsfolk quaked, sobbed and groaned in sympathy, encouraging the Brethren to still greater excesses.

Such scenes were repeated twice by day and once by night with a benefit performance when one of the Brethren died. If the details of the ceremonies are literally as recorded then such extra shows must have been far from exceptional. The public wanted blood and they seem to have got it. Henry of Herford records:

> Each scourge was a kind of stick from which three tails with large knots hung down. Through the knots were thrust iron spikes as sharp as needles which projected about the length of a grain of wheat or sometimes a little more. With such scourges they lashed themselves

> on their naked bodies so that they became swollen and blue, the blood ran down to the ground and bespattered the walls of the churches in which they scourged themselves. Occasionally they drove the spikes so deep into the flesh that they could only be pulled out by a second wrench. . . .

Rules of the Flagellants

The Flagellant Movement, at first at least, was well regulated and sternly disciplined. Any new entrants had to obtain the prior permission of their husband or wife and make full confession of all sins committed since the age of seven. They had to promise to scourge themselves thrice daily for thirty-three days and eight hours, one day for each year of Christ's earthly life, and were required to show that they possessed funds sufficient to provide 4d [pence] for each day of the pilgrimage to meet the cost of food. Absolute obedience was promised to the Master and all the Brethren undertook not to shave, bathe, sleep in a bed, change their clothes or have conversation or other intercourse with a member of the opposite sex.

The entrance fee ensured that the poorest members of society were barred from the Brotherhood; the strict rules, at first at any rate conscientiously observed, kept out the sensation-mongers who wished only to draw attention to themselves or to give unbridled scope to their passions. In these conditions, the public were generally delighted to receive the visits of the Flagellants and, at a small charge, to meet their simple needs. Their arrival was an event in the drab lives of the average German peasant; an occasion for a celebration as well as for the working off of surplus emotion. If the plague was already rife then the visit offered some hope that God might be placated, if it had not yet come then the penance of the Flagellants was a cheap and possibly useful insurance policy. Without at first being overtly anti-clerical the movement gave the villager the satisfaction of seeing his parish priest manifestly playing second fiddle if not actually humiliated. Ecclesiastics had no pre-eminence in the movement; indeed, in theory, they were forbidden to become Masters or to take part in Secret Councils, and the leaders of the movement prided themselves upon their independence from the church establishment.

So bourgeois and respectable, indeed, did the movement at

first appear that a few rich merchants and even nobles joined the pilgrimage. But soon they had reason to doubt their wisdom. As the fervour mounted the messianic pretensions of the Flagellants became more pronounced. They began to claim that the movement must last for thirty-three years and end only with the redemption of Christendom and the arrival of the Millennium. Possessed by such chiliastic convictions they saw themselves more and more, not as mortals suffering to expiate their own sins and humanity's, but as a holy army of Saints. Certain of the Brethren began to claim a measure of supernatural power. It was commonly alleged that the Flagellants could drive out devils, heal the sick and even raise the dead. Some members announced that they had eaten and drunk with Christ or talked with the Virgin. One claimed that he himself had risen from the dead. Rags dipped in the blood they shed were treated as sacred relics. . . .

The Clash with the Church

As this side of the movement's character attracted more attention, so a clash with the Church became inevitable. Already the claim of the Masters to grant absolution from sins infringed one of the Church's most sacred and, incidentally, lucrative prerogative. . . . The German Flagellants took the lead in denouncing the hierarchy of the Catholic Church, ridiculing the sacrament of the eucharist and refusing to revere the host. Cases were heard of Flagellants interrupting religious services, driving priests from their churches and looting ecclesiastical property. . . .

But the turning point came with the declaration of war by the Church. In May, 1348, Pope Clement VI had himself patronised ceremonies involving public flagellation within the precincts of his palace at Avignon but he took fright when he saw that he could not control the movement which he had encouraged. Left to himself he would probably have turned against them sooner, but members of the Sacred College prevailed on him to hold his hand. In mid-1349, the Sorbonne [University] was asked for its opinion and sent to Avignon a Flemish monk, Jean da Fayt, who had studied the phenomenon in his homeland. It seems that his advice was decisive. Shortly after his arrival, on 20 October, 1349, a papal Bull was published and dispatched to the Archbishops. This was followed by personal letters to the Kings of France and England. The Bull denounced the Flagellants for the contempt of Church discipline which they had shown by forming unauthorised associations, writing their own statutes, devising their own uniforms and performing many acts contrary to accepted observances. All prelates were ordered to suppress the pilgrimages and to call on the secular arm to help if it seemed necessary. . . .

[According to Henry of Herford] The Brethren of the Cross 'vanished as suddenly as they had come, like night phantoms or mocking ghosts.' The movement did not die, indeed it was still to be encountered in the fifteenth century, but, as a threat to society or an additional headache to those grappling with the problems of the Black Death, it had effectively ceased to exist.

Useless Fanatics?

It is easy to poke fun at these misguided fanatics. Their superstitions were ridiculous, their practices obscene, their motivation sometimes sinister. But before condemning them one must remember the desperate fear which drove the Flagellants into their excesses. These were men who put themselves to great pain and inconvenience; in part, certainly for the sake of their own souls and their own glory, but in part also in the hope that their sacrifice might induce God to lift from his people the curse that was destroying them. There were few saints among them but, on the whole, they were not bad men. And it is impossible not to feel some sympathy for the person who, when disaster threatens, tries to do something to oppose it, however futile, instead of waiting, in abject despair, for death to strike him down.

They did achieve something. In some at least of the towns they visited they brought about a spiritual regeneration, ephemeral, no doubt, but still real while it lasted. Adulterers confessed their sins, robbers returned stolen goods. They provided some diversion at the places along their route and left behind them a fleeting hope that their pain might bring an end to the greater sufferings of the plague-stricken. But when the Flagellants had passed, often leaving new centres of infection in their wake; when the miracles did not happen, the sick did not recover, the plague did not pass; then the condition of those they left behind them must have been even worse than before they came. On the whole they probably did more harm than good.

One thing at least it is hard to forgive. In his Bull condemning them, Pope Clement VI complained that 'most of them . . . beneath an appearance of piety, set their hands to cruel and impious work, shedding the blood of Jews, whom Christian piety accepts and sustains.' The persecution of the Jews during the Black Death deserves special attention. The part which the Flagellants played in this repugnant chapter was only occasionally of the first importance but it was none the less barbarous for that.

The Search for a Scapegoat

When ignorant men are overwhelmed by forces totally beyond their control and their understanding it is inevitable that they will search for some explanation within their grasp. When they are frightened and badly hurt then they will seek someone on whom they can be revenged. Few doubted that the Black Death was God's will but, by a curious quirk of reasoning, medieval man also concluded that His instruments were to be found on earth and that, if only they could be identified, it was legitimate to destroy them. What was needed, therefore, was a suitable target for the indignation of the people, preferably a minority group, easily identifiable, already unpopular, widely scattered and lacking any powerful protector.

The Jews were not the only candidates as victims. In large areas of Spain the Arabs were suspected of playing some part in the propagation of the plague. All over Europe pilgrims were viewed with the gravest doubts; in June, 1348, a party of Portuguese pilgrims were said to be poisoning wells in Aragon and had to be given a safe conduct to get them home. In Narbonne it was the English who were at one time accused. But it was the leper who most nearly rivalled the Jew as popular scapegoat. . . .

Why Jews Were Targeted

In Germany, and to some extent also in France and Spain, the Jews provided the money-lending class in virtually every city—not so much by their own volition as because they had been progressively barred from all civil and military functions, from owning land or working as artisans. Usury was the only field of economic activity left open to them; an open field, in theory at least, since it was forbidden to the Christian by Canon Law. In cities such as Strasbourg they flourished exceedingly and profited more than most during the economic expansion of the thirteenth century. But the recession of the fourteenth century reduced their prosperity and the increasing role played by the Christian financiers, in particular the Italian bankers, took away from them the cream of the market. In much of Europe the Jew dwindled to a small money-lender and pawnbroker. He acquired a large clientele of petty debtors so that every day more people had cause to wish him out of the way. . . .

But though the economic causes for the persecution of the Jews were certainly important it would be wrong to present them as the only, or even as the principal reason for what now happened. The Jew's role as money-lender predisposed many people to believe any evil which they might hear of him but the belief itself was sincere and had far deeper roots. The image of the Jew as Anti-christ was common currency in the Middle Ages. It seems to have gained force at the time of the First Crusade and the Catholic Church must accept much of the responsibility for its propagation. The vague enormity of such a concept was quickly translated into terms more comprehensible to the masses. In particular the more irresponsible priests spread rumours that the Jews kidnapped and tortured Christian children and desecrated the host. They were represented as demons attendant on Satan, portrayed in drama or in pictures as devils with the beards and horns of a goat, passing their time with pigs, frogs, worms, snakes, scorpions and the horned beasts of the field. Even the lay authorities seemed intent on fostering public belief in the malevolence of the Jews; in 1267, for instance, the Council of Vienna forbade purchases of meat from Jews on the ground that it was likely to be poisoned.

To-day such fantasies seem ludicrous. It is hard to believe that sane men can have accepted them. And yet [historian] Dr Norman Cohn has drawn a revealing parallel between anti-Semitism in the fourteenth century and under the Third Reich [of Nazi Germany]. On 1 May, 1934, *Der Stürmer* devoted a whole issue to alleged murders of Christian children by the Jews; illustrating its text with pictures of rabbis sucking blood from an Aryan child. Most Germans were no doubt revolted by such vicious propaganda but [the concentration camps of] Buchenwald, Auschwitz and Belsen live vividly enough in the memory to save this generation from any offensive sense of superiority to its ancestors. Nor do the still more recent Chinese accusations that American airmen, in 1952, showered the countryside around Kan-Nan Hsien with voles infected with *Pasteurella Pestis*, the bacillus of bubonic plague, suggest that man's infinite capacity for thinking ill of man is in any way on the wane.

Rumors of Well Poisoning

The Black Death concentrated this latent fear and hatred of the Jews into one burning grievance which not only demanded vengeance but offered the tempting extra dividend that, if the Jews could only be eliminated, then the plague for which they were responsible might vanish too. There was really only one charge levelled against the Jews; that, by poisoning the wells of Christian communities, they infected the inhabitants with the plague. . . . Some of the more fanciful reports alleged that the Jews were working under the orders of a conspiratorial network with its headquarters in Toledo [in Spain]; that the poison, in powdered form, was imported in bulk from the Orient, and that the same organisation also occupied itself in forging currencies and murdering Christian children. But these were decorative frills, the attack on the sources of drinking water was the central issue. . . .

A partial explanation may be that many wells in built-up areas were polluted by seepage from nearby sewage pits. The Jews, with their greater understanding of elementary hygiene, preferred to draw their drinking water from open streams, even though these might often be farther from their homes. Such a habit, barely noticed in normal times, would seem intensely suspicious in the event of plague. Why should the Jews shun the wells unless they knew them to be poisoned and how could they have such knowledge unless they had done the poisoning themselves? . . .

There can be little doubt that the majority of those who turned on the Jews believed in the literal truth of the accusations against them. It might be thought that this certainty would have been shaken by the fact that Jews died as fast as Christians; probably faster, indeed, in their crowded and unhealthy ghettoes. . . .

But though such crude suspicions might have been acceptable to the mob, they can hardly have been taken seriously by the intelligent and better educated. Dr [S.] Guerchberg has analysed the attitude of the leading plague tractators. The most remarkable feature is how few references there are to the guilt or innocence of the Jews. Konrade of Megenberg brusquely dismissed the accusations: 'Some say that this was brought about by the Jewish people, but this point of view is untenable.' In his *Buch der Natur* he cites as evidence Jewish mortality in Vienna which was so high that a new cemetery had to be constructed. Gui de Chauliac was equally categoric. Alfonso of Cordova considered that, by all the rules of planetary action, the Black Death should only have lasted a year and that any subsequent extension must be the result of a wicked plot. But he did not specifically accuse the Jews of being responsible. . . .

On the whole, this reticence on the part of the tractators must be taken to indicate that they did not believe the accusations. It is impossible that they did not know what had been suggested and, if they had really thought that a principal cause of the plague was the poisoning of the wells by Jews, then they could hardly have failed to say so in their examination of the subject. Their silence might imply that they thought the idea too ridiculous to mention but it is more likely that they shrank from expressing publicly an unpopular view on an issue over which people were dangerously disturbed.

The Massacres Begin

For it took considerable moral courage to stand up for the Jews in 1348 and 1349 and not many people were prepared to take the risk. The first cases of persecution seem to have taken place in the South of France in the spring of 1348, and, in May, there was a massacre in Provence. Narbonne and Carcassone exterminated their communities with especial thoroughness. But it is possible that the madness might never have spread across Europe if it had not been for the trial at Chillon in September 1348 of Jews said to have poisoned certain wells at Neustadt and the disastrous confessions of guilt which torture tore from the accused. . . .

In most cities the massacres took place when the Black Death was already raging but in some places the mere news that the plague was approaching was enough to inflame the populace. On 14 February, 1349, several weeks before the first cases of infection were reported, two thousand Jews were murdered in Strasbourg; the mob tore the clothes from the backs of the victims on their way to execution in the hope of finding gold concealed in the lining. In part at least because of the anti-Semitism of the Bishop, the Jews of Strasbourg seem to have suffered exceptionally harshly. A contemporary chronicle puts the grand total of the slaughtered at sixteen thousand—half this would be more probable but the Jewish colony was one of the largest of Europe and the higher figure is not totally inconceivable.

From March until July, there was a lull in the persecution. Then the massacre was renewed at Frankfurt-am-Main and, in

August, spread to Mainz and Cologne. In Mainz, records one chronicler, the Jews took the initiative, attacked the Christians and slew two hundred of them. The Christian revenge was terrible—no less than twelve thousand Jews, 'or thereabouts,' in their turn perished. In the North of Germany, Jewish colonies were relatively small, but their insignificance was no protection when the Black Death kindled the hatred of the Christians. In the spring of 1350, those Jews of the Hansa towns who had escaped burning were walled up alive in their houses and left to die of suffocation or starvation. In some cases they were offered the chance to save themselves by renouncing their faith but few availed themselves of the invitation. On the contrary, there were many instances of Jews setting fire to their houses and destroying themselves and their families so as to rob the Christians of their prey.

Why the persecutions died down temporarily in March, 1349, is uncertain. It could be that the heavy losses which the Black Death inflicted on the Jews began to convince all those still capable of objectivity that some other explanation must be found for the spread of the infection. If so, their enlightenment did not last long. But the blame for the renewal of violence must rest predominantly with the Flagellants. It is difficult to be sure whether this was the work of a few fanatics among the leaders or merely another illustration of the fact that mass-hysteria, however generated, is always likely to breed the ugliest forms of violence. In July, 1349, when the Flagellants arrived in procession at Frankfurt, they rushed directly to the Jewish quarter and led the local population in wholesale slaughter. At Brussels the mere news that the Flagellants were approaching was enough to set off a massacre in which, in spite of the efforts of the Duke of Brabant, some six hundred Jews were killed. The Pope condemned the Flagellants for their conduct and the Jews, with good reason, came to regard them as their most dangerous enemies.

Attempts to Protect the Jews

On the whole the rulers of Europe did their best, though often ineffectively, to protect their Jewish subjects. Pope Clement VI in particular behaved with determination and responsibility. Both before and after the trials at Chillon he published Bulls condemning the massacres and calling on Christians to behave with tolerance and restraint. Those who joined in persecution of the Jews were threatened with excommunication. The town-councillors of Cologne were also active in the cause of humanity, but they did no more than incur a snub when they wrote to their colleagues at Strasbourg urging moderation in their dealings with the Jews. The

Believing that the plague was punishment for man's sins, the Flagellants performed self-castigating rituals with the hope of appeasing God.

Emperor Charles IV and Duke Albert of Austria both did their somewhat inadequate best and Ruprecht von der Pfalz took the Jews under his personal protection, though only on receipt of a handsome bribe. His reward was to be called 'Jew-master' by his people and to provoke something close to a revolution.

Not all the magnates were so enlightened. In May, 1349, Landgrave Frederic of Thuringia wrote to the Council of the City of Nordhausen telling them how he had burnt his Jews for the honour of God and advising them to do the same. He seems to have been unique in wholeheartedly supporting the murderers but other great rulers, while virtuously deploring the excesses of their subjects, could not resist the temptation to extract advantage from what was going on. Charles IV offered the Archbishop of Trier the goods of those Jews in Alsace 'who have already been killed or may still be killed' and gave the Margrave of Brandenburg his choice of the best three Jewish houses in Nuremberg, 'when there is next a massacre of the Jews.' A more irresponsible incitement to violence it would be hard to find.

Nor were those rulers who sought to protect the Jews often in a position to do much about it. The patrician rulers of Strasbourg, when they tried to intervene, were overthrown by a combination of mob and rabble-rousing Bishop. The town-council of Erfurt did little better while the city fathers of Trier, when they offered the Jews the chance to return to the city, warned them quite frankly that they could not guarantee their lives or property in case of further rioting. Only Casimir of Poland, said to have been under the influence of his Jewish mistress Esther, seems to have been completely successful in preventing persecution.

The persecution of the Jews waned with the Black Death itself; by 1351 all was over. Save for the horrific circumstances of the plague which provided the incentive and the background, there was nothing unique about the massacres. The Jews had already learned to expect hatred and suspicion and the lesson was not one which they were to have much opportunity to forget. But the massacre was exceptional in its extent and in its ferocity; in both, indeed, it probably had no equal until the twentieth century set new standards for man's inhumanity to man. Coupled with the losses caused by the Black Death itself, it virtually wiped out the Jewish communities in large areas of Europe. In all, sixty large and one hundred and fifty smaller communities are believed to have been exterminated and three hundred and fifty massacres of various dimensions took place. It led to permanent shifts of population, some of which, such as the concentration of Jews in Poland and Lithuania, have survived almost to the present day. It is a curious and somewhat humiliating reflection on human nature that the European, overwhelmed by what was probably the greatest natural calamity ever to strike his continent, reacted by seeking to rival the cruelty of nature in the hideousness of his own man-made atrocities.

The Devastating and Far-Reaching Effects of the Black Death

BY ROBERT S. GOTTFRIED

In the following excerpt from The Black Death: Natural and Human Disaster in Medieval Europe, *historian Robert S. Gottfried outlines the immense effects of the Black Death on the overall makeup of the medieval world. The general psychological response to the massive population loss was Epicurean in nature, or marked by a concern with earthly pleasures. The postplague world also introduced "merchant's time"—an extended workday and a new, urgent understanding of mortality. In addition, Gottfried believes that the failure of Islamic and Christian holy men to offer solace to plague victims resulted in a general dissatisfaction with religious institutions. People began to take their salvation into their own hands; charitable deeds and pilgrimages became popular among Christian laypeople as a means of attaining grace. Meanwhile, the arts revealed a pessimistic mind-set similar to that in the religious world. Images of death and suffering prevailed in both painting and literature. Finally, Gottfried argues that the old social order of the Middle Ages began to crumble after the Black Death. Agricultural workers were in high demand, and the resulting increase in wages caused the nobility to lose their stronghold in society. While the peasants enjoyed a period of economic gain, the aristocracy suffered an identity crisis. Their attempts to deny the good fortune of the lower classes indirectly resulted in the English Peasants' Revolt of 1381—a sign of the general increase in class tensions after the plague.*

By the end of 1351 the Black Death had run its course. . . . There were immediate and pronounced consequences from such enormous population losses, the first and most obvious of which were on human behavior and psychology. The shock was immense, and the mechanics and commonplaces of everyday life simply stopped, at least initially. When plague came, peasants no longer ploughed, merchants closed their shops, and some, if not all, churchmen stopped offering last rites. Many of the responses were described by Boccaccio in *The Decameron*. . . .

The Pursuit of Pleasure

The Epicurean response, perhaps because of a macabre attraction for the dissolute, seems to stand out. Indeed, the pursuit of pleasure is the theme of *The Decameron*. . . . Through all of [the] plague's devastation, the revelers accept and observe the basic tenets and trappings of Christianity, and in many ways follow the rules of good Christian conduct. They are of the upper classes and do not question the hierarchy and order of temporal life. They do not seem to doubt the ultimate truth of the Christian faith and the dogmas of the church. Yet, the accepted Christian trifunctional scheme has changed somewhat. While none of the characters question God's omnipotence, many believe that their future is determined, not by their own actions—one of the tenets of the Roman Church—but by fate, luck, and chance. Good fortune was keenly sought as a sign of divine blessing. Boccaccio's characters valued qualities far different from those esteemed by most of their predecessors. Piety, martial or mechanical skills, scholarly intelligence were not emulated; rather, wit and cleverness were venerated as essential to success. The cuckold, not the cheat, the liar, or the coward, was reviled; the seducer, not the pious schoolman or brave knight, was admired. Rewards and triumph went to the active, to those who helped themselves. The Black Death, at least among a substantial portion of the most vigorous members of society, brought not so much a stoic acceptance of pain and suffering as it did a desire for an active, temporal life.

The Decameron is not the only example of popular literature reflecting the new values. Within a generation, Geoffrey Chaucer's *Canterbury Tales*, which expressed a perspective on life and values similar to that of *The Decameron*, had as much impact on the English reading public as Boccaccio's tales had had in Italy. . . .

Epicureanism seems to have been especially strong among the most influential members of society, particularly the aristocracy and the intelligentsia, and its persistence made for a profound and prolonged moral crisis. Some scholars believe that such a crisis was well underway in 1347 and may have begun with the subsistence economy of the mid-thirteenth century; but all believe that the Black Death, as in so many other concurrent crises, made this one far more pronounced than it had been. Much of the old corporate cooperation and camaraderie were swept away, replaced in many cases by a strong strain of individualism. . . . In the decade after the Black Death, individualism generally was directed towards self-aggrandizement and the pursuit of leisure and pleasure. The collective institutions and the old communality—both rural and urban—so characteristic of the twelfth and thirteenth centuries, were shattered. Old social, religious, and even familial bonds were relaxed. Restoring these bonds was the challenge that faced the people of the late fourteenth and fifteenth centuries.

The Rise of "Merchant's Time"

Among the psychological changes brought by plague was a new sense of time, especially among the bourgeoisie. Traditionally, merchants and churchmen had had a different sense of the temporal. To the churchman, time was infinite, the domain of God. To the merchant, time was finite, a function of distance—for example, the number of days it would take for a ship to sail from Genoa to Bruges—or of changing seasons—for example, the number of days before the Alpine passes would become impassable. Time was money, an attitude that caused considerable consternation among the clergy and prompted theologians to condemn the practice of usury. They argued that usury and all commercial ventures were suspect because they assumed control over the future, a mortgage of time which was reserved for God.

The Black Death changed this; it brought a sense of urgency, especially in urban areas. The work day was extended and night work became common as merchants sought greater profits and workers, higher wages. For example, in Ghent and several other Flemish towns, as the Black Death diminished in late 1349, textile laborers demanded that they be allowed to determine their hours. Clocks and the rhythmic chimes of bells became more important than ever. . . . By the end of the century, [according to Jacques LeGoff in *Time, Work, and Culture in the Middle Ages*] "merchant's time," rather than "the traditional conception of time in Christian theology," became the rule.

The Shortcomings of Islam

Such psychological changes had a pronounced effect on late medieval religion. The depth of both Christian and Muslim religious feeling in the West was one of the foundations of society. Followers of both faiths counted the hereafter more important than life on earth and, given the difficulties and brevity of earthly existence, salvation became paramount. The clergy claimed to be the conduit to salvation and thus held a special position. The Black Death—that sudden, precipitous, painful, and omnipresent killer—intensified the medieval preoccupation with death, judgment, heaven, and hell. Death seemed nearer, and salvation more important, than ever, and the clergy were put to a real test. If, in one fashion or another, they acted responsibly and were able to relieve the anxiety—and, in some cases, hysteria—of their flocks, their position would be strengthened. If not, the faithful would follow another path to heaven.

Generally speaking, both Islamic and Christian holy men failed the test. Muslim theologians, following long-standing doctrine, offered their followers three tenets. First, the faithful should not flee the Black Death, but rather, should stay and accept Allah's will. Second, death by plague was martyrdom, a mercy for the true believer and a punishment for the infidel. Third, in a direct rebuttal of over a thousand years of accepted wisdom, the theologians denied the general medical opinion that plague was a contagious infection transmitted from person to person. They stated that it was foolish to flee from the Black Death because it was God, not men, who disseminated the disease. And there was another reason to reject the advice of doctors; God was good, and contagion was simply incompatible with His very being. . . .

For many of the Islamic faithful, such injunctions must have been all the comfort necessary. But, for others, they were not. The Islamic medical community offered extensive advice, despite condemnation and ridicule by the mullahs and, judging from the popularity of medical works, many Muslims must not have been content with a fatalistic acceptance of God's judgment. Perhaps even more significant as a barometer of popular dissatisfaction with the Islamic clergy—after all, only a small, literate elite was able to afford and read medical treatises—was the spread among the lower classes of a kind of magic designed to ward off plague, or cure those who had already contracted it. The magic consisted mostly of special prayers using numbers or incantations, and talismans and amulets, particularly those made from gold or silver. Hex signs were popular, too, especially when carved in sapphires or ivory. All were considered to be prophylaxes and, as such, were uniformly condemned by the clergy, who claimed that such symbols were an insult to God. But their popularity persisted, another affront and challenge to the authority of the mullahs.

The Shortcomings of Christianity

The damage inflicted on the Christian clergy was far more substantial. In part, this was because the Christian institutional, hierarchical bureaucracy was more extensive than that of Islam; in part, it was because the institutional Christian Church had begun to decline at least two generations before the Black Death, at the time of the pontificate of Boniface VIII (1294–1303). . . . In general, the papacy had grown ever more temporal from the early thirteenth century. It was increasingly concerned with fiscal gain and secular political advantages, and, at the very least, it gave the appearance of neglect in spiritual things. This aspect of papal decline predated the Black Death and was largely extraneous to the effects of depopulation. Further, most historians believe that, as a group, the Avignonese popes were of high caliber. Clement VI, for example, acted coolly and responsibly during the Black Death, waiting himself until the last minute to flee from the plague, and then doing so only on the advice of his doctor. However, when the Black Death brought the crisis of the imperial Christian Church to a head and provided it with as stiff a challenge as it had had since its earliest days, both its spiritual and educational offices were found wanting. Christians did not abandon their faith, but many of them sought alternative paths to spiritual peace and salvation.

The principal failure of the Christian Church was in not providing the necessary solace or support during the crisis. This failure took two forms. First, except for parts of northern Italy, the church supervised the education and licensing of physicians, almost all of whom were clerics. There were surgeons, apothecaries, and nonprofessional practitioners whose training and practice lay outside church jurisdiction, but they had little to do with the theories and tractates about infectious disease that flooded

Europe after 1347. In the long run, virtually all medical advice proved to be useless. As the church would have taken credit had its physicians been able to assuage the pain of plague, so it was to shoulder much of the blame when they failed.

Second and more important, the church did not provide adequate spiritual comfort. Many parish priests fled, leaving no one to offer services, deliver last rites, and comfort the sick. Flight might have been intellectually explicable, but it was morally inexcusable. In the English dioceses of York and Lincoln, close to 20% of the parish priests in certain deaneries fled the Black Death. . . . Given all these circumstances, it is not surprising that many Christians continued to follow their own path to salvation even after the plague had subsided and their priests had returned.

The Upsurge in Charity and Pilgrimages

One direction the faithful took was a strong reinforcement of the traditional idea that works, as well as faith, would help in attaining salvation. . . . One of the most popular of these good works was pious charity, which blossomed after the Black Death and remained popular until the early sixteenth century. In England, about a quarter of all testators' estates, land and movables, went to good works. Hospitals benefited. In France, donations to existing institutions rose about 50% from 1300 to 1350; in England, 70 new foundations were laid between 1350 and 1390. Family chapels were another favorite bequest and, in the fourteenth and fifteenth centuries, there was a marked rise in the number of private masses and chantry priests. This, in turn, was a reflection of the increased popularity of the concept of purgatory, that halfway house where those who would eventually be saved were made to serve time in hell-like conditions in order to purge themselves of sin before they were permitted to enter heaven. Time in purgatory could be shortened by private masses or any number of good works. This charity system of private worship played a large role in late medieval religion and represented a considerable blow to monopoly over church services held by the traditional Christian hierarchy.

Charity was important for another reason. In the dislocation that followed the Black Death, many ecclesiastical institutions could not make ends meet. . . . In some case, pious charity was the only income religious houses had for years after the Black Death. Because of the reduction of population by a third or more, the total sums received might well have been down, and there were some projects—the cathedrals at Siena and Winchester, for example—that could not be finished even with new infusions of charitable bequests. However, the tangible display of good works exerted a strong pull and, at least in England and Italy, per capita bequests were much higher in the 1350s than they had been earlier in the fourteenth century. . . . Charity, especially to hospitals, was an opportunity for a twofold response to the Black Death. It provided for an institution that helped plague victims and it was the kind of good work that counted toward salvation, but could be done directly, without clerical involvement.

Another popular "good work" was pilgrimage to a religious shrine. Here, too, the faithful were performing a religious act directly, using a saint rather than a priest to intercede in their behalf. Pilgrimage could be undertaken to a place of the first rank, such as Rome, Jerusalem, or St. James Compostela in Galicia, or to one of innumerable local shrines that contained relics or anything else of religious importance. . . . Pilgrimage was not an easy undertaking. Given the deplorable condition of late medieval roads and the threat of brigands and pirates, it was a very dangerous task. As such, it was considered one of the most important good works and counted a great deal toward salvation. Wills from England and Italy showed a marked upsurge in the number of bequests for pilgrims and pilgrimages. In the 1350s and 1360s, there was a glut of travel guidebooks—some sober and earnest, others fabricated and sensational—which describe the process of pilgrimage and tell the pilgrim where to stop to eat and spend the night, and even the proper way to venerate particular saints. Sir John Mandeville's *Travels,* written in 1357, was the most popular of these guides. . . . In many ways, pilgrimage represents all the changes, good and bad, which the Black Death brought to religion—the passion to do good deeds, the strong conviction that this would help to attain salvation, and the decision to take part of one's fate in one's own hands, combined with the frivolity and lightheartedness so evident in Chaucer's brightest pilgrim, the Wife of Bath. . . .

Ties to the Protestant Reformation?

The idealized selfless image of the Christian clergy suffered during and after the Black Death. Many people believed, often unjustly, that the clergy were greedy, self-centered, and filled with a sense of their own importance. It must be stressed, however, that while confidence in the institutional church waned, faith in Christianity itself did not; rather, the imminence of death, brought closer than ever by plague, made the need for salvation more pressing. One consequence was the spread of mysticism and lay piety. Mystics, the most famous of whom were Meister Eckhart, John Ruysbroek, John Tauler, and Henry Suso, believed that God lived in every individual and that His presence was felt in proportion to one's ability to suppress intrinsic material and sensual inclinations and subject one's will to that of God. Obedience, self-denial, and prayer were crucial. Lay piety expressed itself in organizations such as the Brethren of Common Life, formed in the Netherlands in the late fourteenth century. Aside from their profound sincerity, the most striking characteristic of mysticism and lay piety was the lack of need for a formal clergy to lead the way to paradise. Many postplague Christians felt they could communicate directly with God.

It is tempting to tie the decline of the institutional church to the Protestant Reformation of the sixteenth century, as many historians of the nineteenth and twentieth centuries have done. Such a direct connection, spanning close to 200 years, is a bit presumptuous. The Christian Church had many problems before the advent of the second plague pandemic. It was a huge, unwieldy, and vastly complicated institution, and, even at its nadir, there was much about it that was good. But the Black Death made an issue of the proper function of the clergy. It made people ever more conscious of the omnipotence of God and the inevitability of Judgment Day. A poorly behaved clergy made many people wonder about alternative means of salvation. Perhaps the best link between the Black Death, the decline of Holy Mother Church, and the Protestant Reformation of the sixteenth century was the growing role of indulgences in the fourteenth and fifteenth centuries. Attuned to the anxieties of their flock, churchmen redoubled the emphasis on good works. From the 1350s, apparently on papal orders, new stress was put on indulgences, or grants of time off from purgatory bestowed by the church, which drew on what it termed a "treasury of merits," or good deeds accumulated from Christ, the patristic fathers, and saints. Indulgences were not given freely, but usually in anticipation of a gift of money; always mindful of turning a profit, church leaders began to sell them in increasing numbers to a richer public. While indulgences were not the only thing that spurred Martin Luther, their use and sale inspired him to nail up his 95 Theses. . . .

Pessimism and Death in the Arts

Much of the cruelty and violence, as well as the piety and joy, of the late fourteenth and fifteenth centuries can be understood only by keeping in mind the new omnipresence of plague and the possibility of sudden, painful death. In the High Middle Ages, an era of expansion and fruition, literature and art expressed a buoyant optimism. After the Black Death, this was replaced by a pervasive pessimism. In addition to the carefree tone of some of the works of Boccaccio, Chaucer, and [François] Villon, a sense of melancholy entered the literature. . . .

People were fascinated by death. Preachers advised people to go to sleep every night as if it were their last and as if their beds were their tombs. The frailty of human life and the brevity of earthly glory were stressed. Ascetic mediation was best because everyone ended up as dust and worms. Putrefaction was evidence of sin; only saints' bodies did not decompose. In the Early and High Middle Ages, people accepted the inevitability of death and prepared for it, but they were rarely preoccupied with it. Burials were often in common graves, and elaborate tombs were rather rare. The Black Death changed all this. Funerals became festivals, the greatest event of a lifetime. . . . After the plague, funerary monuments and death masks became common, and their themes changed. Many brasses showed shrouded, macabre corpses or skeletons with snakes and serpents surrounding and protruding from their bones; on their faces were grisly, toothy smiles. Tombs in the Netherlands showed hideous images of naked corpses with tightly clenched hands, rigid feet, gaping mouths, and bowels filled with worms. In Germany, woodcuts called "The Art of Dying" appeared. They were linked panels showing the drama of death. Death was painful—in contrast to the peaceful slumber of years past—and people were to tremble at its coming. All this marked the appearance of the *ars moriendi,* the cadaver and death as a major motif in art and literature.

Perhaps the best examples of this preoccupation with the motifs of death and despair were in the fine arts. . . . Tastes were changing. The new patrons were more conservative and had doubts about their materialism, their goals, and sometimes even their success, in view of the new preoccupation with salvation.

They were becoming increasingly guilty and introspective.

There were changes in the artists as well as the patrons. The Black Death killed many individual artists and, in some cases, entire schools or guilds of painters, sculptors, and masons who had

Images of death and suffering prevailed in postplague art and literature. This illustration depicts the Black Death as a demon.

been inspired by similar themes and hence worked together. This not only eliminated some of Europe's greatest masters, but it made it very difficult to train and develop new talent. A good example comes from England, where thirteenth-century guilds of artists produced superb miniatures. The Black Death killed off such a great number of these artists that no new masters appeared, and the English could not maintain their standards in this special art form.

The effects of new patrons and artists were visible immediately. Preplague Tuscan art was warm and sympathetic. It stressed personal relationships and, when it dealt with religious themes, emphasized the humility of Jesus, the Virgin Mary, and the saints. Postplague art, like postplague thought and funerary monuments, was obsessed with the most gruesome aspects of pain, and with the image of death. This new image can be seen in many forms, one of the best examples being Francesco Triani's great fresco, "The Triumph of Death," in the Camposanto in Pisa, painted around 1350. Death is shown, not as an airy skeleton, as was usually the case before the Black Death, but rather, as a horrible old woman cloaked in black, with wild, snakelike hair, bulging eyes, clawed feet with talons, and a scythe to collect her victims, whom she feeds to snakes and toads. Death was like a bird of prey, sweeping down on its victims. . . .

There were also changes in Northern art. The disasters of the fourteenth century did not cause a lessening of creativity, but rather, a change in direction. The old patrons of the preplague era were generally great churchmen, especially bishops and abbots. The new patrons, mostly members of the bourgeoisie, were less learned and sophisticated, and had more somber tastes. The art they sponsored no longer showed the harmony between man, reason, and nature—the forms of God placed in their proper, natural hierarchy and popularized in the twelfth and thirteenth centuries. Rather, the new art was narrative: it told stories, sometimes of God, but more frequently of secular things. Themes of earthly escapism were common. For better or worse, as bourgeois patrons became more important, artists were no longer auxiliaries of priests.

A Somber Turn in Literature

As in the fine arts, the Black Death brought changes in literary styles and ideas, most of which were somber. A good example is the later works of Boccaccio. *The Decameron,* his literary masterpiece, was written in the vernacular and was enormously popular. Its cynicism reflected a common perception during and immediately after the Black Death. But his attitudes soon changed. While *The Decameron* is guilt-free, Boccaccio's later works show a more sober side. *The Corbaccio,* written in 1354–55, is gloomy, pessimistic, truculent, and ascetic, attitudes that hardened even more as Boccaccio grew older and began to think about his own salvation. . . .

Plague brought similar changes to the literature of northern Europe. The brevity of life, the folly of exuberance, and the painful horror of death by plague were emphasized along with the more hedonistic themes of Chaucer. The purveyors of gloom and doom stressed that death did have at least one thing to recommend it—it brought equality to all the social orders. Exegesis became popular, but commentators offered little solace from the Bible. The Old Testament was drawn on more heavily than before, and it was stressed that God afflicted his chosen people as well as his enemies with plague. The most heavily quoted book from the New Testament was Revelations, where plague was described as God's punishment for the sins of man; in all, man no longer seemed to be God's favorite creature. Generally, the works of Chaucer and Villon notwithstanding, themes of youth, exuberance, happiness, and joy were played down. The dance of death became a common literary motif. Mystery plays with religious themes also became common, and they usually told of human decay and the torments of hell. There was much written about the ages of life, mostly in the form of calendars, with analogies drawn to the seasons of the year. In the thirteenth and early fourteenth centuries, the calendars emphasized spring and summer; in the late fourteenth and fifteenth centuries, they turned to themes of autumn and winter.

Benefits for the Peasants

There were similar changes in the conceptions and realities of social order. The trifunctional system of the eleventh and twelfth centuries began to come apart around 1250, but it was the Black Death, with its massive depopulation, that finally brought the old order down. The plight of the clergy has already been described. In their role as intermediaries with God, they failed to provide solace to plague victims, and the medical education system, built around them, failed to provide physical comfort. In a world in which performance of an appointed role was very important, many clerics no longer seemed to be doing their jobs.

The Black Death brought crisis to the nobility. The loss of between a quarter and a half of the West's population ended Europe's subsistence crisis. Depopulation meant widespread mobility and fluidity for those not tied to the land. The value of agricultural products began to fall, and it stayed low relative to that of industrial goods until the sixteenth century; at the same time, depopulation made agricultural workers scarce and, thus, much more valuable. Wages rose rapidly. . . .

At Cuxham Manor in England, a ploughman who was paid 2 shillings a week in 1347 received 7 shillings in 1349, and 10 shillings, 6 pence, by 1350. The result was a dramatic rise in standards of living for those in the lower part of the third order. . . . The economic and social ramifications of this new relationship between wages and prices were far-reaching. But the social consequences were great, too, and were felt immediately after the Black Death. For the peasants who farmed the land, depopulation, providing they survived the plague, was a great boon. Yet for those who held the land as lords—the aristocracy and the clergy—it was disastrous.

At first, the landed classes tried to reinstate the trifunctional order through legislation issued by the representative bodies that they controlled. Authorities all across Europe promptly began to enact sumptuary laws. In France, a 1349 Statute of Labor attempted to limit wages to pre-1348 levels. This failed, and two years later a new law was enacted which allowed for a 33% rise. In 1349, the King's Council in England passed an Ordinance of Labour, that froze wages. This was followed in 1351 by a Parliamentary Statute of Labourers that tried to do the same thing. Parliament, it should be remembered, consisted almost entirely of men from the first two orders, with a few great merchants from the third, most of whom owned property in their own right. . . . All of these efforts were for naught, and landlords discovered that the only way to keep laborers was to pay the going rate. . . .

Identity Crisis for the Aristocracy

The position of the nobility was affected by the Black Death in other ways. Plague did not honor social class, and mortality among the nobility approximated that of the general population.

Since inheritance was more important to them than it was to the peasants, the biological crisis was more severe. This made an already bad situation worse. Given the high infant and child mortality rates of the Middle Ages—combined, they probably killed three of four children before age ten—and maternal mortality of about 20%, it was hard in the best of circumstances to produce an heir. In England, 75% of all noble families failed to produce a male heir through two generations. This meant continual flux among the aristocracy as old families died out and new ones replaced them.

One response to this tremendous fluidity was a renewed emphasis by the older families on the importance of knightly ritual. In the Early and, to a slightly lesser extent, the High Middle Ages, the aristocracy had been warriors. In the Late Middle Ages, while they were still soldiers, their military ascendancy was being challenged by infantry men using new weapons. Consequently, late medieval aristocrats became more conscious than ever of their role as heavy cavalry, turned with disdain upon the baseborn footmen who as often as not defeated them on the battlefield, and immersed themselves in elaborate rituals of chivalry. Armor became plated rather than chain-linked. Tournaments, formerly held for practice and profit and usually conducted with the actual weapons of war, became dress celebrations, with combatants using blunted swords and lances. . . . However obsolete they might become, such events helped the old aristocracy preserve their identity. A similar, albeit somewhat gentler, retreat for the lords was to manners and courtesy. Dozens of books were compiled on nurture and grace, and how to dress, act, eat, and think like a gentleman. The nobility were developing a contempt for manual labor and laborers, and even for merchants and the most profitable commercial activities.

High mortality among the aristocracy, then, was more damaging than it was among the peasantry because of the importance to the lords of proper patterns of inheritance. At the same time, there was a new balance between prices and wages. Land was no longer as valuable as it had been, but laborers were worth much more than before. Society was changing from a labor-intensive base to a land-intensive one, and markets for foodstuffs collapsed as population decreased. As more tenants died, lords had to hire wage laborers to farm their lands, and these workers demanded better pay and conditions. On the manors of the Clare estates in

England, for example, reaping wages per quarter averaged less than five pence per acre between 1340 and 1349. In 1349, wages doubled. Estate auditors did their best to reduce their labor costs, sometimes by arbitrarily slicing them in half. But, generally, such efforts were ineffective. Drawing an example from England again, aristocratic incomes fell over 20% between 1347 and 1353.

The Rise in Popular Rebellion

One of the most important effects of the Black Death was its role in the provocation of popular rebellion. Plague alone did not cause the many rebellions of the fourteenth century; rather, they had a long and complex history. But until the late thirteenth century, rebellions in Europe were comparatively few and generally of religious inspiration. . . . These later uprisings first took on political overtones and then, exacerbated by the depopulation of the Black Death, became increasingly socioeconomic in nature.

The revolts after the Black Death had several common characteristics. First, they took place during a general breakdown in law and order. Both the court and police systems in most of fourteenth century Europe were operated by local landholders. King and church might claim a superior jurisdiction, but, with the exception of parts of England and Italy, law enforcement was a local prerogative. As the economic and military power and social prestige of the landholders declined, their ability to uphold the law also diminished. Coupled with the increasing tendency toward violence, this resulted in an enormous upswing in crime. In England, where royal power was probably more extensive than elsewhere in Europe, the incidence of homicide from 1349 to 1369 was about twice that of the period 1320 to 1340, despite the general drop in population. As the fabric and structure of society deteriorated, people resorted increasingly to violence to settle their differences.

A second plague-related reason for the spate of revolts was a heightened sense of class of identity. This was especially so among the peasantry, on whose labor and produce the other classes depended. . . . They believed that their interests conflicted with those of the first two orders, the clergy and the nobility. This sense, perhaps first evident in the early fourteenth century as the effects of successive famines began to cripple the manorial system, was made even more apparent after the Black Death when the lords refused to recognize the peasants' changed situation.

The new relationship between wages and prices was a third reason. Boccaccio claimed that prices shot up during the Black Death. He was right, but by 1351 things had changed. Prices for industrial goods remained high, a reflection of increased demand and a shortage of skilled laborers to make special products. But population was reduced so much that, as soon as seeds were sown and crops harvested, the great subsistence crisis ended. It took another year or two for distribution systems to return to normal; then food prices began to fall. Because of the population decline, wages went up, and so did standards of living. The Late Middle Ages has with good reason been called the "Golden Age of Laborers," and many scholars believe that real wages were higher in the fifteenth century than at any time in history until the twentieth century. As discussed, members of the first two orders chafed at the rising wage level and tried to reverse this trend through legislation. To members of the third order, this was the cruelest blow of all. Having at last attained some economic security in a market economy, they were now being asked to live within new, artificial restraints.

Three major revolts resulted from postplague social and economic tensions. Two—the *Jacquerie* in France and the Peasants' Revolt in England—were peasant uprisings, and the other—that of the *Ciompi* in Florence—was an urban-industrial rebellion. . . .

The English Peasants' Revolt

The English Peasants' Revolt of 1381 is the best known of the postplague uprisings. The immediate cause of the revolt was a series of poll, or head, taxes assessed three times between 1377 and 1381. As with the *Jacquerie* and the *Ciompi,* most of the conditions that set off the English revolt had been festering before 1347, but plague accelerated the changes and general social tensions. The peasants wanted to preserve the high wages and new mobility brought by depopulation, and the lords, threatened on all fronts, wished to maintain the status quo despite the new economic conditions.

The revolt erupted in eastern England, the richest part of the kingdom, rather than in the depressed northern and western counties. It began when a number of peasants in Essex refused to pay the head tax and drove the collectors from their village. Spontaneously, peasants and townsmen rose against what they perceived to be injustices. The thrust of the rebellion, which was led by a wealthy peasant named Wat Tyler and an unemployed priest named John Ball, was antinoble, anticlerical, and antiauthoritarian. Tyler, leading a peasant army on London, exhorted his followers to "kill all lawyers and servants of the king." According to Froissart, Ball claimed: "Ah, ye good people, the matter goes not well to pass in England, nor shall not do so till everything be common, and that we may be all united together and that the lords be no greater masters than we be. What have we deserved or why should we be thus kept in serfdom; we be all come from one father and one mother, Adam and Eve." . . .

The English Peasants' Revolt followed the patterns of the *Jacquerie,* the *Ciompi,* and other postplague revolts. After an early period of success, bloody revenge on their tormentors, and random violence—among other things, the archbishop of Canterbury was beheaded and most of the brothels in greater London were destroyed—the aristocracy and gentry regained their dominance and, as in France, brutally put down the rebels. But, in England, as on the Continent, the rebels came away with substantial gains. Poll taxes were eliminated and there were no more statutes or ordinances fixing wages or limiting mobility. The peasants benefited from high wages and, by 1400, the old bonds of villeinage had loosened or crumbled. There were no more peasant revolts in England in the Late Middle Ages because the peasants had no reason to revolt.

Both the urban and the rural disorders reflected the sharp class conflicts that developed after the Black Death. A general social discontent arose, not so much over conditions in postplague Europe as over the ruling classes' attempts to deny the lower order the better fortune depopulation brought. The breakdown of the old trifunctional system and the increasing violence made revolt easier and more palatable. The revolts were spontaneous and poorly organized and, in time, were easily put down. Perhaps their most telling legacy was the end of the traditional hierarchical social structure, which was replaced by interclass bitterness and tension. How might the changes affecting the generation which survived the Black Death be characterized? First, immediate depopulation of at least a third ended, in a single blow, Europe's subsistence crisis. By the early 1340s, most Europeans were getting poorer. They held and farmed less land, produced and, hence, consumed less food, and had a smaller say in determining their future than at any time since the Dark Ages. By contrast, the landlord class was becoming increasingly powerful. Europe was evolving, as were parts of Asia and Africa, into an impoverished "agrocentric" society. By 1350, this process was dramatically reversed. High wages and low prices were ruining the lords, while members of the third order were embarking on 150 years of comparative prosperity. But in the short term, for those who survived the first plague, the psychological effects of the Black Death were far more important. People were traumatized. They lost faith in their own abilities, in the old values, and if not in God then in the traditional ways in which He had been propitiated. Europe was plunged into a moral crisis. The old order was collapsing and the new one was not yet in place.

Advances in Medicine, Surgery, and Public Sanitation

By Anna Montgomery Campbell

Anna Montgomery Campbell was a medieval historian and the author of The Black Death and Men of Learning, *an important book first published in 1931 about the effects of the plague on various intellectual fields. In the following excerpt, Campbell discusses the advances made in medicine, surgery, and public sanitation as a result of the mass mortality of the plague. Although it was once banned by the Church as a sacrilegious practice, the pope and other public officials finally allowed for the dissection of human corpses during the Black Death. Many believed that a knowledge of anatomy would help to prevent and cure infectious diseases. According to Campbell, surgery in general became a legitimate medical practice after the Black Death, and many towns saw the rise of surgeons' guilds to protect their medical interests. In addition, the fear of infection incited by the plague gave rise to legislation concerning public sanitation; Italian cities were the first to practice quarantines and to place embargoes on incoming peoples and goods. Finally, the medical tracts or tractates written during the Black Death furthered the general public's knowledge of personal hygiene and prophylaxis. Although some of the theories in these tracts were far off the mark, Campbell stresses that the Black Death caused a beneficial dialogue about the nature of contagion and infection that greatly benefited the progress of science and medicine.*

In addition to the plague tractates evoked by the Black Death from medical men of learning, there were other consequences of the great mortality for them and their profession.

First we may inquire as to the number of physicians and surgeons who were among its victims—a question presenting peculiar difficulties because of the scarcity and irregularity of records kept while the epidemic was at its height. . . .

The Survival Rate of Physicians

We can reasonably assume the death of only eleven [of the well-known medical treatise writers] as a result of the Black Death, and most of these are based on date of death without definite statement of cause. Twelve of the physicians and surgeons who wrote on the pestilence in the second half of the century, besides those whose names have already been used in computations of mortality, were practicing at the time of the Black Death. The same is true of twenty-four from whom we have no writings on the pestilence, so that of fifty-one men who are known to have been practicing physicians when the epidemic occurred, eleven died when it was prevalent in their communities, four soon after it had passed and before it had entirely ceased, and thirty-six survived.

It is gratifying to one who has acquired respect and liking for these earnest, and often gifted, scholars of a past age to note that while, as far as we know, only two of the writers of plague tractates died, two—Dionysius Colle and Guy of Chauliac—tell us that they contracted the plague in 1348 and, though their recovery was despaired of, were cured by their own treatment. In view of the deadliness of the disease . . . this record would seem to be a feather in the cap of fourteenth-century medicine.

A noticeable point about the lives of many of the physicians is the number of instances in which the Black Death seems to have formed a turning point, or at least to have played an important part, in their careers. [Arab physician and writer] Ibn al-Khatīb, on the death by pestilence in 1348 of Ibn al Gaijâb, wezîr of Sultan Jusuf of Granada, succeeded him in that office and remained wezîr till a few years before his death in 1374. John Dondi [a European physician and writer, as the names to follow] became personal physician of Charles IV in 1349, and in the same year Gallus of Strahov appears to have found a place at the court of the same monarch. In 1348 John of Tornamira began his practice at Montpellier, where he remained till 1369; and John Arderne states that from the year of the first pestilence, 1349, till 1370 he lived at Newark, in the county of Nottingham. Guy of Chauliac was papal physician at Avignon from 1348 till his death in 1368, and the debut of another surgeon, John of Parma, as a member of the papal staff in 1348 was due to the death by pestilence of a papal surgeon, Peter Augerii.

A New Esteem for Surgery

The fact that the last three of these men were surgeons (two of them, Guy of Chauliac and John Arderne, very distinguished ones), brings to attention a momentous change that occurred in the field of medicine in the fourteenth century, and which may have been accelerated by the Black Death. Till about the end of the thirteenth and the beginning of the fourteenth centuries, surgery had never been held in esteem, nor, except for a period of brilliance in Roman times, had it developed beyond a rudimentary stage. As intellectual activity and curiosity increased in Europe, and the dependence of accurate knowledge upon experimentation became more evident, enterprising spirits began to break through tradition and prejudice. In the thirteenth century the practice of dissection was begun by Mundinus of Bologna; he wrote an *Anatomy* which was the first work of the medieval period devoted entirely to the subject. His pupil and successor was Bertruccius, who, in turn, was instructor of Guy of Chauliac. Others who blazed the trail were John Gaddesden, Lanfranc, Henry of Mondeville, and John Yperman, the last three authors of works on surgery in the early fourteenth century.

But the *Great Surgery* of Guy of Chauliac and the *Practice* of John Arderne show a decided advance over the earlier works of the century, and from the middle of the century a double tendency is increasingly discernible: to break down the barriers between physicians and surgeons, and to standardize the profession of surgery. For instance, in Venice it was against the law for surgeons to practice medicine, and 31 December, 1348, a fine was inflicted upon Andreas of Padua because, in the mortality of 1348, he went out as a physician and cured over a hundred of those smitten with the pestilence; this the Venetian judges attributed to accident rather than to wisdom. Yet in September, 1349, Nicholas of Ferrara was invited to Venice from Padua and praised for his knowledge and experience in surgery: Both of the cities in which he had practiced, it is said, had profited by his knowledge of diseases of the lower parts of the body [*crepatorum*], which required a specialist. Guy of Chauliac, though really a surgeon, was physician of Clement VI; and John of Parma, starting in 1348 as papal surgeon, later became physician and surgeon. . . .

In England an indication of the trend toward raising surgical standards was the formation of surgeons' gilds[i.e. guilds]. The Oxford gild was incorporated by Dr. Northwood, chancellor of the university, in 1348; it included barbers and waferers [physicians], but lectures in surgery began to be given about 1350. In 1362 it was decreed that every London surgeon must belong to a gild; further regulations were enacted twice later in the century, and before its close gilds of surgeons had been founded in half a dozen other towns.

The first judicial post-mortem was held in 1302 at the University of Bologna, where Mundinus had earlier inaugurated the practice of anatomical dissection; post-mortems occurred also in other universities, Gentile of Foligno holding one publicly in 1341 at Padua. But the first recorded instances of public officials, other than educational authorities, decreeing autopsies seem to occur during the Black Death. In a letter written from Avignon by a cleric of the Netherlands, the statement is made that

> anatomical dissections have been performed by physicians in many states of Italy, and also in Avignon, by order and command of the pope, that the origin of this disease might be known, and many bodies of the dead have been opened and dissected.

One city in Italy which took such a step was Florence, for in a list of public expenditures dated 30 June, 1348, there is an item "for giving more corpses to physicians who requested them, in order to be able to learn more clearly the diseases of the bodies." Public dissections were decreed by several other cities during the century. In 1376 the duke of Anjou, governor of Languedoc, issued an ordinance to the officers of justice in that province: They were to suppress in their districts the illegal practice of medicine, and to send every year to the medical faculty of Montpellier for dissection the corpse of an executed criminal. The purpose of both decrees is set forth: It is an attempt to save what mortality, epidemics, and war have left of the population; dissection is desirable, since experience is mistress of affairs, and darts foreseen usually cause less damage, and visible dangers can more easily be avoided, than the hidden and the unknown. The last part of the fourteenth and the early fifteenth centuries was a period of renown for the medical faculty of Montpellier, for it then combined the knowledge of surgery with that of medicine. . . .

The Birth of Public Sanitation

Public sanitation is connected, at least indirectly, with medical affairs, and was decidedly affected by the Black Death. Public measures to combat the plague began in Italian cities in 1348, and developed by the seventies into the quarantine, which dates from this time, and by 1402 into the establishment of a lazaretto [hospital] by Venice. This city, on a Sunday in March, 1348, elected a commission of three to suggest plans for the public safety. On the following Thursday they made recommendations to the Great Council, at a meeting open to all officials and judges, for measures to prevent the corruption of the city, dealing chiefly with the disposal of bodies. The problem became so acute that the city purchased ships at a high price and carried the dead to islands. In Florence, where the ravages of the plague were so terrible that it is sometimes called "the Florentine pestilence," the *gonfaloniere de giustizia* [chiefs of justice] and the twelve *buoni uomini* [bondsmen] chose eight of the wisest and most respected citizens, whose names are preserved in the city records, to exercise a sort of dictatorship. This was the first office of public health . . . in the history of the republic; its duties were to see to the removal of decaying matter and infected persons from the city, and to supervise the markets. The priors [secular authorities who represented the gilds] of Florence at this time took into their service physicians to study the nature of the disease and suggest to them remedies to adopt. . . .

Embargoes of various kinds, to keep out of the city infected persons or goods, were set up in 1348. As early as 13 January of that year Lucca issued a decree that no Genoese or Catalan, or anyone who had within the past year been in any city or parts of Romagna, should dare enter the city of Lucca, under penalty of the property and person of him who should violate the decree. The Great Council of Venice forbade the bringing of a sick foreigner into the city on pain of imprisonment, burning the ship, or other penalty, and as the pestilence advanced cities prohibited entry to all from without, with the result that merchants could not travel from place to place. Matters finally reached such a point, Guy of Chauliac tells us, that guards were kept in the cities and villages to prevent the entry of anyone who was not well known. A document of Pavia, dated 1387, is preserved, providing for the payment of seven officials who had been deputed to stand at the city gates for that purpose. . . .

A Step in the Right Direction

It seems that, as wars are the diet upon which military establishments thrive, so the Black Death was, on the whole, of benefit to the medical profession. The loss of leaders in any field of intellectual activity is a blow effects of which are incalculable, and physicians and surgeons of outstanding ability and achievement are numbered in the twenty-five or thirty per cent which our statistics indicate as the approximate mortality among medical practitioners. But it is undeniable that in the last half of the fourteenth century there was a strikingly large number of distinguished physicians and surgeons, and that they were much sought after by rulers and other persons of eminence. Their salaries and fees rose perceptibly, other evidence bearing out [historian A.] Palmieri's observation that the medical art was very lucrative in the second half of the fourteenth century, with physicians of the seventies and eighties, in the region he was studying, leaving large patrimonies, and being the richest proprietors of certain districts which he mentions.

The practice of post-mortems and of anatomical dissection in general appears to have been stimulated by the epidemic, and the upward movement of surgery to have been accelerated. The public was freely instructed, in tractates generously written and scattered abroad, in principles of hygiene, sanitation, and personal prophylaxis. . . . It was an era of public sanitary ordinances, and marked the beginning of formal quarantine. There was intensive study of the causes of epidemic, leading to many conclusions which are today considered false, such as the relation between heavenly bodies, natural phenomena, and human beings; but also giving a decided impetus to development of the theories of contagion and infection. While some of the hypotheses advanced, such as infection by the glance of the eye, were wide of the mark, others, for instance the development of immunity, and carriers of disease, have been more happily borne out by the researches of succeeding scientists.

Competing Historical Interpretations of the Black Death

By Faye Marie Getz

Faye Marie Getz is a historian most known for her works on medieval medicine. In the following excerpt from her article that appeared in the Journal of the History of Biology, *Getz discusses the changing historical interpretations of the Black Death over time. She believes that, with the publication in 1832 of* The Black Death in the Fourteenth Century *by German epidemiologist Justin Hecker, the "gothic" interpretation of the plague gained popularity. In this school of thought, it was common for scholars to take a romantic view of the disaster. The Black Death was seen as a mysterious phenomenon with a silver lining, a disaster that ushered in a new age of rapid social change, eventually resulting in the high culture of the Renaissance. With the advent of the 1960s, however, this interpretation shifted. Scholars at this time began to argue that social and intellectual institutions endured the plague relatively unchanged. With this new criticism of "gothic epidemiology," economic, social, and art historians argued that the Black Death did not incite rapid changes; rather, they argued that many of the changes historians had focused on as a consequence of the plague were actually present before the disaster. Getz concludes that the Black Death's place in history actually resides somewhere in the middle; along with a gradual change in social behavior and institutions was humanity's amazing ability to endure hardship and begin again.*

Faye Marie Getz, "Black Death and the Silver Lining: Meaning, Continuity, and Revolutionary Change in Histories of Medieval Plague," *Journal of the History of Biology*, Summer 1991, pp. 265–89.

Humanity can rest assured that no disaster, however terrible, is without redeeming social content. At least this is what one supermarket tabloid [*National Enquirer*, May 6, 1986] would have us believe. A headline in point: "Even though 55 million died, Black Death that Wiped out Europe Had a Good Side!" The accompanying article began, "The horrifying Black Death wiped out more than 55 million people in Europe during the Middle Ages—but the catastrophe changed the world forever by giving birth to the Renaissance." According to an interview with a much-published medievalist, famous authorities agree that the Black Death not only reduced Europe's population from 75 million to 20 million, but it also put an end to those dismal Middle Ages and "nurtured geniuses like Michelangelo and Leonardo da Vinci." The piece ends with the comforting observation that this awesome slaughter indeed had "a silver lining." One shudders to contemplate the death of 55 million people as a recipe for social renewal. But the thesis that the article presents, wedged though it may be between advertisements for secret good luck charms and miracle diet pills, has an undeniable appeal. History, like a novel, ought to make sense, and how could the death of most of the population of Europe have happened for no good reason?

Competing Interpretations of the Black Death

It is undeniable that interest in various disasters ebbs and flows with the times. . . . But interest in the Black Death seems almost perennial. The subject must possess the largest bibliography of any single epidemic in history, and it is the only disease whose six hundredth anniversary (in 1948) has been celebrated with the delivery and publication of papers in tribute to its awesome power. No microorganism has ever achieved this level of superstardom.

The Great Plague of the fourteenth century provoked written comment from those who lived through it, and after it, as had no other natural event in the medieval West. This amount of documentation for the pandemic has spawned a Black Death industry, which began during the early nineteenth century alongside the science of epidemiology itself. Early nineteenth-century German scientists in particular not only saw in medieval accounts of the Black Death a laboratory for their own speculation on nature and epidemic disease, but also saw evidence for man's progress toward a new age, the Renaissance, which was born after a horrible cataclysm just as the world was renewed after the Great Flood. This Romantic—or, as will be argued, gothic—construction of the Black Death remains a powerful theme in the history of epidemiology today. But since at least the 1960s another, competing interpretation of the Black Death has been offered, especially by historians of the *Annales* school [that attempts to look at history from a wide range of perspectives]. Rather than being impressed by the apocalyptic accounts of medieval plague chroniclers, these historians, using for the most part municipal records, see in the Black Death evidence for the enduring nature of medieval social and intellectual institutions. The gothic interpretation of the Black Death comprises themes of teleology, individual heroism, abrupt change, death, and, most notably, a dialectic between opposing forces. The *annalistes* find in the plague continuity, the ordinary, gradual change, and the collective experience of large groups of everyday people simply getting on with their lives. Both points of view show how a changing cultural climate affects the way in which historical evidence is interpreted, and a study of these contrasting interpretations of the Black Death throws light on many of the unspoken assumptions of historical and scientific writing. . . .

Hecker and the Rise of Epidemiology

Discussions of the meaning and impact of the medieval plague are rare in learned sources after 1500 and before the nineteenth century. Enlightenment thinkers attached little importance to the fourteenth-century plague. The *Encyclopédie* [French Enlightenment Encyclopedia] of 1757 devotes scarcely a quarter column to the Black Death, which is described as dwarfing previous epidemics, as compared to nearly two columns on the plague of Athens in 431 B.C. A turning point for Black Death scholarship came, however, with the great cholera epidemic of the early 1830s, and it was accompanied by the rise of epidemiology in Germany.

Most notable among the early epidemiologists to interpret the Black Death was Justin Hecker, whose *Der schwarze Tod im viersehnten Jahrhundert* (The Black Death in the Fourteenth Century) was published in 1832. Hecker was a physician at the new Friedrich Wilhelm University in Berlin, and was the author of a popular history of medicine. His book on plague obviously struck a responsive chord in nineteenth-century society. It spread at lightning speed into English by 1833, and subsequently into Italian, Dutch, and French. By 1834 the English version was in its second edition, and it was last printed in 1972. An edition for the Sydenham Society was prepared in 1844, and this was reprinted several times along with others of Hecker's shorter works.

The history of epidemics was for Hecker what class struggle was to be for [authors of *The Communist Manifesto*, Karl] Marx and [Friedrich] Engels: a hitherto unnoticed force for historical change and progress. Epidemics, he argued, had a much greater influence on the course of world history than did wars or politics. And yet, their study was still in its infancy. In an address to German physicians that was often printed with *Der schwarze Tod,* Hecker asserted that French and English universities refused to devote themselves to the history of epidemiology. It was therefore the special calling of German universities to remedy this omission.

Ours is a new age, Hecker told the German physicians, and the history of epidemiology is a science worthy of it. . . . We cannot possibly understand the diseases of our own time unless we compare them with those of the past, he argued, and only the history of epidemics will allow us to do that. We must study "the diseases of nations, and of the whole human race," and not just individual outbreaks, so that we can comprehend the whole of nature herself. Hecker declared that

> the very stones have a language, and the inscriptions are yet legible which, before the creation of man, were engraved by organic life on eternal tablets. . . . Epidemics leave no corporeal traces, whence their history is perhaps more intellectual than the science of the Geologist, who, on his side, possesses the advantage of dealing with subjects which strike the senses. . . .

[According to Hecker] the magnitude of the crisis [of the Black Death] was beyond the grasp of puny mortals: . . .

> These revolutions . . . are performed in vast cycles, which the spirit of man, limited as it is, to a narrow circle of perception, is unable to explore. . . . By annihilations they awaken new life, and when the tumult above and below the earth is past, nature is renovated, and the mind awakens from torpor and depression to the consciousness of an intellectual existence.

The Start of "Gothic Epidemiology"

In the end, then, conflict was resolved and all was for the best. But the new age did not begin at once. The Black Death excited humanity to extremes of behavior: "Unbridled demoniacal passions" unfolded side by side with the noblest and most courageous behavior. In Hecker's words, "all that exists in man, whether good or evil, is rendered conspicuous by the presence of great danger." What is more, after the plague "a greater fecundity in women was everywhere remarkable," which proved, to Hecker at least, "the prevalence of a higher power in the direction of general organic life."

Hecker's contribution to the history of epidemiology went beyond the Romantic notion that the synthesis of science and history should ultimately take place; he also gave to the medieval plague the legacy it retains even today by defining the Black Death's social impact. Not surprisingly, perhaps, he fixed on two of the most bizarre and repellent phenomena of the late Middle Ages: the processions of Flagellants who roamed from town to town scourging themselves and prophesying the end of the world, and the hideous pogroms that were launched against more than three hundred Jewish communities, whose members were said to have brought about the plague by poisoning the wells. In the case of the persecution of the Jews, Hecker flew in the face of medieval opinion and insisted that the Jews, rather than being part of the cause of the plague as some medieval people suggested, instead were part of its effect or social impact: They became the indirect victims of plague rather than its cause. So potent were Hecker's definitions of these twin social impacts that the pair dominate medieval Black Death scholarship even today.

Brought together in Hecker's treatise are what might be called the seeds of subsequent histories of the fourteenth-century plague, which are only now beginning to dissipate. First is the idea that the fourteenth-century plague marked the beginning of a new age, a break with a decadent past and the ushering in of a more vigorous time. The second is the concept of the Black Death as a natural phenomenon somehow beyond human comprehension: terrible, seductive, and marvelous. It was an event in nature and human history like no other, in which man was but an insignificant pawn on Nature's chessboard. The third is the idea that the Black Death was, like Charles Dickens's French Revolution [in his novel *A Tale of Two Cities*], both the best and the worst of times. The Black Death represented a war of opposites, whose result was the Renaissance. The fourth is Hecker's emphasis on the disease's decisive social impact, especially on the more bizarre and morbid aspects.

I would like to refer to this particular group of factors by the term "gothic epidemiology," in the sense of "gothic" used by literary historians, because it would appear that the essential elements of what is usually called the gothic sensibility are present in Hecker's writings. The term "gothic" was first applied to medieval Teutonic and Germanic tribes, but by the eighteenth century it had come to mean almost anything that offended Enlightenment sensibilities, especially the so-called gothic architecture from the late Middle Ages. Modern readers are most familiar with how the English Romantic poets reveled in what might be called essential elements of gothic sensibility: an interest in distant and exotic places and times, especially in the Middle Ages and the Orient; the celebration of the power of nature and the ineffability of nature's essence; the unity of disparate elements—of good and evil, the hideous and the beautiful, the dead and the living; the seduction of the primitive and wild in nature, of the bizarre; the insignificance of human beings against nature; the existence of geniuses; the importance of individual experience; and finally the emphasis on suffering, death, and redemption. . . .

The Breaking of a Tradition

Hecker's fame as the founder of Black Death epidemiology has enjoyed an unbroken tradition. Alfonso Corradi, professor of pathology at the University of Palermo, credited him with this in his monumental statistical survey *Annali delle epidemie* [*Annals of the Epidemics*]. In 1893, in his history of German epidemiology, the great epidemiologist August Hirsch called Hecker one of the first and greatest of his kind, and he was joined in this by [medieval historians] Max Neuburger and Julius Pagel. Nearer to our own time Philip Ziegler, Stuart Jenks, and Nancy Siriasi have all drawn attention to Hecker's pioneering effort.

This is not to say that the historians mentioned above can be thought of as enthusiastic proponents of gothic epidemiology. Indeed, some might argue that Hecker as collector of useful facts must be separated from Hecker as weird epidemiologist—but such separations have seldom been made. If Hecker and his gothic epidemiology seem to us, now, a curious aberration, a detour along the road to us, they did not seem so to previous generations. . . .

The attack on gothic epidemiology began in earnest after World War II, and paralleled the tendency among historians to fall away from condemnations of the Middle Ages as those bad old days. Critics of the notion that the Black Death marked the birth of a new age and a significant break with the past have in general focused their attention much less on the immediate social impact of plague and much more on what historians have called the *longue durée*—that is, the resiliency—of medieval institutions, which enabled them to endure repeated crises and indeed to recover from them. . . .

New Interpretations of the Plague's Effects

A desire to chronicle (and to admire) the continuing experience of ordinary people has helped lead economic and social historians to attack the notion that the plague caused a huge and sudden drop in population, which led to crop failures, labor shortages, the end of feudalism, and the growth of a mercantile culture. Many have noticed that the plague returned time and again in what are usually called "echo epidemics," and it should not be thought of as having had a sudden impact on population. . . .

Historical demographers have questioned the sudden-population-drop thesis by demonstrating that the population of western Europe in general, and of England in particular, was already falling by the time the Black Death struck. Scholars like M.M. Postan have suggested that more attention ought to be paid to declining economic prosperity at the end of the thirteenth century and to the effects of the Europe-wide famine of 1314–1317. If there was a demographic crisis of the second half of the fourteenth century that continued into the fifteenth, many have argued, then it began some thirty years before 1348 and endured longer than could be accounted for by plague alone. . . .

Art historians have adopted many aspects of gothic epidemiology to explain the cause of what was seen as a profound shift, especially in Italian painting, after the medieval plague. . . . [In 1951, in *Painting in Florence and Siena After the Black Death*, Millard] Meiss saw in post-plague art and literature a kind of Cold War between medievalism and modernism, and he attributed to plague the late-medieval obsession with macabre themes. Against this thesis, Joseph Polzer [in his article, "Aspects of the Fourteenth-Century Iconography of Death and the Plague"] argued that the medieval obsession with death and decay preceded 1348. He used for his example the gigantic Triumph of Death paintings in the Campo Santo in Pisa, which were models for many later depictions of the *dance macabre* or "dance of death" paintings of the fifteenth century. The Triumph of Death, with its yawning graves and rotting corpses, has been offered as a prime example of the effect of the Black Death on Italian painting. But Polzer demonstrated that the Triumph preceded, rather than followed, the plague in Pisa. . . .

The most concerted assault on gothic epidemiology has come from historians of *mentalités* [mentalities]. These historians are interested in recovering and exploring historical modes of thought, and in the case of the impact of the Black Death, they have tended to refrain from gothic interpretations—preferring instead to emphasize the enduring nature of medieval modes of thought and institutions, and to minimize the more horrific aspects of plague as being transitory rather than pivotal. Notable among these historians of *mentalités* is Robert Lerner, who questioned whether the Black Death caused medieval people to think that the end of the world and the Last Judgment were at hand. Lerner noted [in his article, "The Black Death and Western European Eschatological Mentalities"] a number of prophecies that did indeed predict that such a time was coming—but he also pointed out that these prophecies were old ones that preceded the fourteenth-century plague and that were . . . made to fit the situation at hand. . . .

One of the most effective assaults on the gothic mentality with regard to the Black Death has come from French historian Élisabeth Carpentier [in her article, "Orvieto: Institutional Stability and Moral Change" in Bowsky, ed., *The Black Death: A Turning Point in History?*]. With what might be called an Enlightenment stoicism, Carpentier called the plague "one catastrophe amid others." . . . In her study of the plague at Orvieto she used municipal records, and not apocalyptic medieval chronicles, in order to conclude: "we have not found any revolutionary change or permanent destruction in the different sectors that we have studied so far." With regard to economic stability, she said, "the plague aggravated a preexisting situation; it did not cause a profound change" [page 108]. Religious life, rather than assuming bizarre aspects as Hecker had suggested, continued on as before. "The epidemic," Carpentier concluded, "attacked individuals and not their institutions" [page 113]. . . .

The tension between the advocates of the Black Death as the herald of a new age, and those who see plague as proof of the resiliency of medieval mentalities, is rapidly dissolving. The conflict/resolution model, with its overtones of teleology [and] progress . . . is proving less useful to historians of epidemiology than one emphasizing continuity, gradual change, and the stoicism of the ordinary person. Historians of the plague are gravitating more and more to an intensive study of the local impact of the Black Death. Such local studies reveal diversity—in economic and demographic impact, in the availability of historical sources, and in the interpretation these sources allow. The Black Death still retains its "silver lining," but even that is changing: from proof of the awesome power of nature to level mankind and transform history, to proof of humanity's ability to endure even the worst crisis, to rebuild, and to start again.

FOR FURTHER RESEARCH

General Studies of the Black Death

Norman F. Cantor, *In the Wake of the Plague: The Black Death and the World It Made.* New York: Free Press, 2001.

Phyllis Corzine, *The Black Death.* San Diego: Lucent Books, 1997.

George Deaux, *The Black Death, 1347.* London: Hamish Hamilton, 1969.

Francis Aidan Gasquet, *The Black Death of 1348 and 1349.* New York: AMS Press, 1980.

Robert S. Gottfried, *The Black Death: Natural and Human Disaster in Medieval Europe.* New York: Free Press, 1983.

Geoffrey Marks, *The Medieval Plague: The Black Death of the Middle Ages.* Garden City, NY: Doubleday, 1971.

Johannes Nohl, *The Black Death: A Chronicle of the Plague.* Trans. C.H. Clarke. New York: Harper & Row, 1969.

Terence Ranger and Paul Slack, eds., *Epidemics and Ideas: Essays on the Historical Perception of Persistence.* Cambridge, England: Cambridge University Press, 1992.

Philip Ziegler, *The Black Death.* New York: Harper & Row, 1969.

Localized Studies of the Black Death

Samuel J. Cohn Jr., *The Cult of Remembrance and the Black Death: Six Renaissance Cities in Central Italy.* Baltimore, MD: Johns Hopkins University Press, 1992.

Michael W. Dols, *The Black Death in the Middle East.* Princeton, NJ: Princeton University Press, 1977.

Mark Ormrod and Philip Lindley, eds., *The Black Death in England.* Stamford, CT: Paul Watkins, 1996.

Colin Platt, *King Death: The Black Death and Its Aftermath in Late-Medieval England.* Toronto, Canada: University of Toronto Press, 1997.

J.F.D. Shrewsbury, *A History of the Bubonic Plague in the British Isles.* Cambridge, England: Cambridge University Press, 1970.

Social, Economic, and Psychological Effects of the Black Death

William M. Bowsky, ed., *The Black Death: A Turning Point in History?* New York: Holt, Rinehart, and Winston, 1971.

Anna Montgomery Campbell, *The Black Death and Men of Learning.* New York: AMS Press, 1966.

David Herlihy, *The Black Death and the Transformation of the West.* Ed. Samuel K. Cohn Jr. Cambridge, MA: Harvard University Press, 1997.

Mavis E. Mate, *Daughters, Wives, and Widows After the Black Death: Women in Sussex, 1350–1535.* Rochester, NY: Boydell & Brewster Press, 1998.

Millard Meiss, *Painting in Florence and Siena After the Black Death.* Princeton, NJ: Princeton University Press, 1957.

Daniel Williman, ed., *The Black Death: The Impact of the Fourteenth Century Plague.* Binghamton, NY: Center for Medieval and Renaissance Studies, 1982.

Biological and Medical Perspectives on the Black Death

Leonard Fabian Hirst, *The Conquest of Plague: A Study of the Evolution of Epidemiology.* Oxford, England: Clarendon Press, 1953.

William H. McNeill, *Plagues and Peoples.* Garden City, NY: Anchor Press, 1976.

Susan Scott and Christopher Duncan, *Biology of Plagues: Evidence from Historical Populations.* Cambridge, England: Cambridge University Press, 2001.

Graham Twigg, *The Black Death: A Biological Reappraisal.* New York: Schocken Books, 1984.

Primary Sources and Literary Perspectives on the Black Death

Leon Bernard and Theodore B. Hodges, eds., *Readings in European History.* New York: Macmillan, 1958.

Giovanni Boccaccio, *The Decameron.* Trans. J.M. Rigg. London: Navarre Society, 1921.

Geoffrey Chaucer, *The Canterbury Tales.* New York: Henry Holt, 1928.

Rosemary Horrox, ed., *The Black Death.* Manchester, England: Manchester University Press, 1994.

William Langland, *Piers Plowman.* London: E. Arnold, 1978.

James B. Ross and Mary M. McLaughlin, eds., *The Portable Medieval Reader.* New York: Viking Press, 1949.

General Studies of the Middle Ages and Plague Outbreaks

John Aberth, *From the Brink of the Apocalypse: Confronting Famine, War, Plague, and Death in the Later Middle Ages.* New York: Routledge, 2000.

David Levine, *At the Dawn of Modernity: Biology, Culture, and Material Life in Europe After the Year 1000.* Berkeley: University of California Press, 2001.

Andrew Nikiforuk, *The Fourth Horseman: A Short History of Epidemics, Plagues, Famine, and Other Scourges.* New York: M. Evans, 1991.

Barbara W. Tuchman, *A Distant Mirror: The Calamitous 14th Century.* New York: Alfred A. Knopf, 1978.

Periodical Articles on the Black Death

C.S. Bartsocas, "Two Fourteenth Century Greek Descriptions of the Black Death," *Journal of the History of Medicine,* vol. 21, 1966.

Norman F. Cantor, "Studying the Black Death," *Chronicle of Higher Education,* April 27, 2001.

David E. Davis, "The Scarcity of Rats and the Black Death: An Ecological History," *Journal of Interdisciplinary History*, Winter 1986.

Michael W. Dols, "The Comparative Communal Responses to the Black Death in Muslim and Christian Societies," *Viator*, vol. 5, 1974.

Stephen R. Ell, "Interhuman Transmission of Medieval Plague," *Bulletin of the History of Medicine,* vol. 54, 1980.

Tony Godfrey-Smith, "Plague and the Decline of Medieval Europe: Correlation or Coincidence?" *Australian National University Historical Journal I,* 1964.

Ken Lastufka, "Bohemia During the Medieval Black Death: A Pocket of Immunity," *East European Quarterly,* September 1985.

Robert E. Lerner, "The Black Death and Western European Eschatological Mentalities," *American Historical Review,* June 1981.

Charles L. Mee Jr., "How a Mysterious Disease Laid Low Europe's Masses," *Smithsonian,* February 1990.

John Norris, "East or West: The Geographic Origin of the Black Death," *Bulletin of the History of Medicine,* vol. 51, 1977.

J.W. Thompson, "The Aftermath of the Black Death and the Aftermath of the Great War," *American Journal of Sociology,* vol. 26, 1920–1921.

INDEX

agriculture, 12–13
AIDS, 28, 31
Albert (duke of Austria), 68
Alfonso of Cordova, 64–65
anthrax, 27–29
antibiotics, 10
Arabs, 62
Arderne, John, 89, 90
aristocracy, 82–84
arts
  pessimism and death represented in, 78–80
  postplague, 100–101
astrology, importance of, 48–49

Ball, John, 86
Bertruccius, 90
Bible, 81
Black Death
  AIDS and, 31
  anthrax as cause of, 27–29
  arrival of, in Europe, 18–19
  decline of Europe and, 11–12
  economic consequences of, 27, 43–46, 82–87
  effect of, on religion, 73–78
  Epicurean response to, 71–72
  funeral practices during, 37–39, 78
  gothic interpretation of, 96–99
  historical interpretations of, 13–14, 94–102
  impact of, in England, 41–46
  influence of, on the arts, 78–81, 100–101
  Jews as scapegoat for, 62–69
  misery caused by, 39–40
  panic incited by, 20–22
  precautions taken against, 34, 49–50
  Protestant Reformation and, 76–78
  psychological changes brought about by, 25, 71–73, 87
  as punishment from God, 56
  reactions to, 35–36
  rebellions after, 84–87
  scientific research on, 29–30
  social consequences of, 13, 81–87
  spread of, 12
    role of trade in, 11
  treatment of sick during, 37
  see also bubonic plague; plague
Black Death, The (Ziegler), 13
bloodletting, 51–53
Boccaccio, Giovanni, 33, 80–81, 85
Boniface III (pope), 22
Boniface VIII (pope), 74
Bovine Spongiform Encephalitis. *See* mad cow disease
Bradwardine, Thomas, 44
Brethren of Common Life, 77
Brethren of the Cross. *See* Flagellant movement
bubonic plague, 24
  as cause of Black Death, 26–27
  causes of, 10–11, 16–18, 30
  first recorded incidence of, 10–11

modern threat of, 30–31
outbreaks of, 10–11, 18–19
spread of, 11, 27
symptoms of, 19–20
see also Black Death; plague

Campbell, Anna Montgomery, 88
Canterbury Tales (Chaucer), 71, 76
Cantor, Norman F., 26
Carpentier, Élisabeth, 101
Catholic Church
Flagellant movement and, 60–61
lack of priests in, 42, 44
persecution of Jews and, 63
shortcomings of, 74–75
cattle disease. *See* anthrax
charity, 75–76
Charles IV (emperor), 68
Chaucer, Geoffrey, 71, 76, 81
China, 18
Christianity
faith in, 76–77
shortcomings of, 74–75
Ciompi, 85
Clement VI (pope), 60, 61–62, 66, 74, 90
clergy
death among, 42, 44
decline in importance of, 76–78, 81–82
failure of, 73–75
Cohn, Norman, 63
commerce, 11
Corbaccio, The (Boccaccio), 81
Corradi, Alfonso, 99
crime, 84

"Dance of Death" rituals, 19
dead, burial of, 38–39
death
fascination with, 78–80
portrayal of, in the arts, 78–80
Decameron, The (Boccaccio), 33–40, 71, 80–81
depopulation, effect of, 81–82, 86–87, 100
dissections, 90–91
Dondi, John, 89

earthquakes, 42
Eckhart, Meister, 77
economy, 43–46, 82–87
England
English Peasants' Revolt, 85–87
impact of Black Death in, 41–46
plague in, 27
English Peasants' Revolt, 85–87
Epidemics and Ideas (Slack), 13
epidemiology
gothic, 98–99
rise of, 96–97
Europe
decline in, before Black Death, 11–12, 13
medieval society of, 12–13
population growth in, 12
psychic damage to, 25, 71–73, 87
social decline in, 13

feudal system, 12–13
Flagellant movement, the, 98
claims made by, 60
clash with church by, 60–61
medieval psyche and, 55
origins of, 55–56
persecution of Jews and, 61–62, 66
purpose of, 61
ritual of, 57–59
rules of, 59–60
fleas, 16–18, 27, 30
Florence, 22
Black Death in, 33–40
funerals, 37–39, 78

Gabriele de Mussis, 19
Gaddesden, John, 90
Gallus of Strahov, 89
genome project, 31
Getz, Faye Marie, 94
Gottfried, Robert S., 13, 70
Great Plague of London, 11
Great Surgery (Guy de Chauliac), 90
Guy de Chauliac, 64, 89–90, 92

Hecker, Justin, 96–99
Henry of Mondeville, 90
Herlihy, David, 28
Hippocrates, 48, 53
Hirsch, August, 99
History of the Peloponnesian War (Thucydides), 11
hoof and mouth disease, 28

Iceland, 29
India, 18
individualism, 72
indulgences, 77–78
infectious disease
  resistant to antibiotics, 31
  role of, in natural balance, 10, 12
  see also plague
inheritance, 83
Islam, 73–74
Italy, 20–22

Jacquerie, 85
Jenks, Stuart, 99
Jewish persecution
  effects of, 68–69
  efforts to stop, 66–68
  Flagellant movement and, 61–62, 66
  Hecker on, 98
  massacres and, 65–66
  reasons behind, 62–63
  well poisoning and, 64–65

John of Burgundy, 47
John of Parma, 90–91
John of Tornamira, 89

Karlsson, Gunnar, 29
al-Khatīb, Ibn, 89
Knighton, Henry, 41
Konrade of Megenberg, 64

laborers
  increased wages for, 82, 85
  shortage of, 44–46, 82, 85
Landgrave Frederic of Thuringia, 68
Lanfranc, 90
lay piety, 77
LeGoff, Jacques, 72–73
lepers, 62
Lerner, Robert, 101
literature
  pessimism and death represented in, 78, 80–81
Little Ice Age, 13
Luther, Martin, 78

mad cow disease, 28
Mandeville, John, 76
McNeill, William H., 10
medicine
  advances in, 88–93
  esteem for surgery in, 90–92
  importance of astrology to, 48–49
  Islamic, 73–74
  medieval, 22–23
  modern, 10
  progress of, 53
  public sanitation and, 92–93
medieval society, 12–13
Mee, Charles L., Jr., 16
Messina, 18–19, 20
Milan, 22
Mundinus of Bologna, 90, 91

mysticism, 77

Neuburger, Max, 99
New Testament, 81
Nicholas of Ferrara, 90
nobility, 82–84

O'Brien, Stephen J., 31
Old Testament, 81
overpopulation, 12

Pagel, Julius, 99
peasants
  benefits for, 81–82
  revolts among, 84–87
physicians, 89–90
pilgrimage, 76
Pistoia, 22
plague
  bloodletting as remedy for, 51–53
  cause of, 30
  different forms of, 24–25
  medieval theories about, 22–23, 47–48
  modern threat of, 30–31
  preventive measures against, 23–24, 49–50
  psychic damage caused by, 25, 71–73, 87
  spread of, 27, 34–35
    role of commerce in, 11
  symptoms of, 34
  transmission of, 24
  well poisoning as cause of, 64–65
  see also Black Death; bubonic plague
Plague of Justinian, 11
Plagues and Peoples (McNeill), 10
pleasure, pursuit of, 71–72
pneumonic plague, 24
Polzer, Joseph, 100–101
Postan, M.M., 100
Practice (Arderne), 90
priests
  failure of, 74–75
  lack of, 42, 44
Protestant Reformation, 76–78
purgatory, 75

rats, 16–18, 27, 29
rebellion, popular, 84–87
religion
  charity and, 75–76
  effect of Black Death on, 73–78
  Muslims and, 73–74
  pilgrimage and, 76
  Protestant Reformation and, 76–78
  see also Catholic Church; Christianity
Renaissance, the, 96
Revelations, 81
rinderpest, 28
"Ring Around the Rosy" (nursery rhyme), 21–22
Roman Empire, 11, 18
Ruprecht von der Pfalz, 68
Ruysbroek, John, 77

sanitation, 92–93
Scots, 44
septocemic plague, 24–25
sheep, 43
Siena, 22
Siriasi, Nancy, 99
Slack, Paul, 13
social order, 81–87
St. Joseph's Hospital and Medical Center, 29–30
surgery, 90–92
Suso, Henry, 77

Tauler, John, 77
Thompson, Edward I., 28–29

Thucydides, 11
Time, Work, and Culture in the Middle Ages (LeGoff), 72–73
trade routes, 13
Travels (Mandeville), 76
Triani, Francesco, 80
"Triumph of Death, The" (Triani), 80, 101
Twigg, Graham, 27, 28
Tyler, Wat, 86

values, 71–72
Venice, 22

wells, 64–65
World War II, 29

Xenopsylla cheopis, 16–17

Yersinia pestis bacillus
  as cause of plague, 11, 16–18
  effect of, in humans, 19–20
  in fleas, 30
  spread of, 27
Yperman, John, 90

Ziegler, Philip, 13, 55, 99